James Blake Bailey

The Diary of a Resurrectionist 1811-1812,

to which are added an account of the resurrection men in London and a short

history of the passing of the anatomy act

James Blake Bailey

The Diary of a Resurrectionist 1811-1812,
to which are added an account of the resurrection men in London and a short history of the passing of the anatomy act

ISBN/EAN: 9783337012632

Printed in Europe, USA, Canada, Australia, Japan

Cover: Foto ©ninafisch / pixelio.de

More available books at **www.hansebooks.com**

"THE DISSECTING ROOM." By Rowlandson.

THE DIARY

OF

A RESURRECTIONIST

1811–1812

TO WHICH ARE ADDED AN ACCOUNT OF

THE RESURRECTION MEN IN LONDON

AND A SHORT HISTORY OF THE PASSING OF
THE ANATOMY ACT

BY

JAMES BLAKE BAILEY, B.A.

LIBRARIAN OF THE ROYAL COLLEGE OF SURGEONS OF ENGLAND

LONDON

SWAN SONNENSCHEIN & CO., Lim.

PATERNOSTER SQUARE

1896

INTRODUCTION

THE "Diary of a Resurrectionist" here reprinted is only of a fragmentary character. It is, however, unique in being an actual record of the doings of one gang of the resurrection-men in London. Many persons have expressed a wish that so interesting a document should be published; permission having been obtained to print the Diary, an endeavour has been made to gratify this wish. To make the reprint more interesting, and to explain some of the allusions in the Diary, an account of the resurrection-men in London, and a short history of the events which preceded the passing of the Anatomy Act, have been prepared.

The great crimes of Burke and Hare drew especial attention to body-snatching in

Edinburgh, and consequently there have been published ample accounts of the resurrection-men in Scotland.* For this reason, Edinburgh has been omitted from the present work.

As to the genuineness of the Diary there can be no doubt. It was presented to the Royal College of Surgeons of England by the late Sir Thomas Longmore. In his early days, Sir Thomas was dresser to Bransby Cooper, and assisted him in writing the *Life of Sir Astley Cooper*.

At the suggestion of Lord Abinger, it was decided to introduce an account of the resurrection-men into the book. The information for this was partly obtained by Mr. Longmore from personal communication with some of the resurrection-men, who were then living in London. One of these handed over portions of a Diary he had kept during his resurrectionist days. This was preserved for some years at Netley, and was afterwards presented to the

---

* See *Sketch of the Life of Robert Knox*, by HENRY LONSDALE (London, 1870); and *The History of Burke and Hare and of the Resurrectionist Times*, by GEORGE MACGREGOR (Glasgow, 1884).

College, as stated above. A few extracts from the Diary were printed in the *Life of Sir Astley Cooper*.

The information respecting the resurrection-men is very scattered; the two most useful works for getting up this subject are the *Life of Astley Cooper* before mentioned, and the *Report of the Committee on Anatomy* published in 1828. Most of the detailed information has to be sought for in the newspapers of the period. The accounts there given are, however, generally of such an exaggerated character that it is often very difficult to arrive at the truth. When any fresh scandal had given prominence to the doings of the resurrection-men, the newspapers saw " Burking" in every trivial case of assault. If a child were lost, the paragraph announcing the fact was headed, "Another supposed case of Burking." Reports of the most ridiculous character were duly chronicled as facts by the newspapers of the day. Sometimes over a hundred bodies were supposed to have been found in some building, and it was expected that several persons of eminence would be named in the subsequent proceedings. Search in the papers nearly always fails to find any further mention of the case.

In reading these accounts it must be remembered that " Burking " did not always mean killing a person for the purpose of selling the body, but it referred to the mode adopted by Burke and Hare in killing their victims, viz., suffocation. Elizabeth Ross is called a " Burker," and may be found so described in Haydn's *Dictionary of Dates*. She murdered an old woman named Catherine Walsh, but in the report of her trial there is no evidence of her having attempted to sell the body.

The broadside here printed is an excellent example of this exaggeration. The facts are so circumstantial, that it appears as though there could be no mistake. Enquiry at Edinburgh, however, shows that no such case occurred. Mr. A. D. Veitch, of the Justiciary Office, has very kindly made search, and can find no record of Wilson's supposed crimes. Had the statements in the broadside been true, there is no doubt that this case would have been referred to in books on Medical Jurisprudence. Poisoning by inhalation of arsenic is rare, and Wilson's would have been a leading case. There would also have been great opportunities for studying *post mortem* appearances, as it is stated that three bodies were found in Wilson's possession. Search through the chief books on the subject has failed in finding any reference whatever to this case.

"BURKING BY MEANS OF SNUFF.

" *The following Account is of so serious a Nature that no one can be too cautious how they receive Snuff from Strangers.*

" It appears that, on Monday se'nnight, a man, named John Wilson, was apprehended at Edinburgh on a charge of Burking a number of persons by introducing arsenic into snuff kept by him. He had long excited the suspicion of the police of that place, but so deep-laid were his diabolical schemes that he eluded their vigilance for a considerable time, until Monday last. When, on the moors, on that day, between Lauder and Dalkeith, practising his dreadful trade, it appears that the victim of Wilson's villainy was a poor man travelling over the moor, whom he accosted, and offered a pinch of snuff. He took it, and it had the desired effect. The next individual whom he accosted was a labouring-man breaking stones, who was asked the number of miles to Edinburgh; when answered, he then offered his snuff-box to the labourer, which was refused, alleging that he never used any. Wilson urged him again, which excited the man's suspicions, but he took the snuff, and wrapped it up in paper, and carried it to a chemist at Dalkeith, who analysed it, when it proved to be mixed with arsenic. The police were then informed of Wilson's villainies, who went in pursuit of him, and after a search of him for several days was at length apprehended at a place three miles from Edinburgh, driving rapidly in a vehicle like a hearse, which, on examination, contained three dead bodies. They were recognised from their dresses to be an elderly man, and his wife and son, who were seen travelling towards Lauder the day before.

"Wilson was immediately ironed and conveyed to Edinburgh, and a sheriff's inquest was held on the bodies. After an investigation of nearly two hours a verdict of Wilful Murder was returned against John Wilson, who was fully committed to the Calton gaol to take his trial at the ensuing sessions.

"Wilson is described as a desperate character, and of ferocious countenance. He is supposed to have been two or three years in this abominable practice, and to have realised a considerable sum in the course of that time. His career is now stopped, and that justice and doom which overtook a Burke and a Hare are his last and only portion.

"LINES ON THE OCCASION.

Of Burke and of Hare we have heard much about,
Yet Burking's a trade that was lately found out—
Their plans of despatching were wicked indeed,
'T was thought of all others that theirs did exceed ;
But the scheme first invented of Burking by snuff,
May yet be prevented by taking the huff,
For if strangers invite you to take of their dust,
Decline their kind offers—refuse them you must ;
And would you be safe, and keep from all evil,
Shun them as pests as you'd shun the d——l ;
By these means you'll live, avoiding all strife,
Shunning snuff takers all the days of your life.

"Printed for the Publishers by T. KAY."

The difficulty of getting reliable information is increased by the incomplete nature of most of the newspaper records. In many cases there is an account of a preliminary examination of some of the men who were arrested for body-stealing. The report states that they were remanded, but further search fails to find any subsequent notice of the case. It is often impossible to fix who the men where who thus got into trouble, as they nearly always gave false names : unless they were too well-known to the police who arrested them, they invariably did this.

For the photographs, from which the illustrations of the house at Crail are taken, the writer is indebted to the kindness of Prof. Chiene, of Edinburgh.

THE

DIARY OF A RESURRECTIONIST

CHAPTER I.

THE complaint as to the scarcity of bodies
for dissection is as old as the history of
anatomy itself. Great respect for the body of
the dead has characterised mankind in nearly
all ages ; *post mortem* dissection was looked
upon as a great indignity by the relatives of
the deceased, and every precaution was taken
to prevent its occurrence.

It would be beyond the scope of the present
work to attempt a history of anatomical teaching ; as will be pointed out later on, the
resurrection-men did not come into existence
until the early part of the eighteenth century.

In Great Britain the study of medicine and
surgery was much hampered at this date by the

scarcity of opportunities by which the student might get a practical acquaintance with the anatomy of the human body. A knowledge of anatomy was insisted upon by the Corporation of Surgeons, as each student had to produce a certificate of having attended at least two courses of dissection. It is unnecessary to point out the wisdom of this condition in the case of men who were to go out into the world as surgeons, and, consequently, to have the lives of their fellow-men in their hands. The attendance on the two courses of dissection could be evaded, and this was frequently done. The Apothecaries' Hall had no such restriction, and, consequently, many men went thither and received a qualification to practise, although they were quite unacquainted with human anatomy. The work of such 'prentice hands one trembles to think of ; whatever experience these men did gain was obtained after they began to practise, and so must have been at the expense of their patients, who were generally those of the poorer class in life.

It was pointed out by Mr. Guthrie, that in the then state of the law a surgeon might be punished in one Court for want of skill, and in another Court the same individual might also be punished for trying to obtain that skill. Before the Anatomy Committee, in 1828, Sir Astley Cooper narrated the case of a young man who was rejected at the College of Surgeons on account of his ignorance of the parts of the body; it was found, on enquiry, that he was a most diligent student, and that his ignorance arose entirely from his being unable to procure that which was necessary for carrying on this part of his education.

When bodies were obtained for dissection it was generally by surreptitious means; the newly-made grave was too often the source from whence the supply was obtained. At first there was no direct trade or traffic in subjects by men who devoted all their efforts to this mode of obtaining a livelihood. The students supplied their own wants as they arose. Mr. G. S. Patterson told the Committee that at St. George's Hospital the students had to exhume bodies for their own use.

In the *Diary of a Late Physician* Samuel Warren has given us a chapter on this subject, which he calls "Grave Doings," and which is

probably founded on fact. The object in the
expedition here recorded was, however, rather
to obtain a valuable pathological specimen,
than to get a body for dissection. Writers of
fiction have made use of body-snatching, and
have given a gruesome turn to their stories by
making the body, when uncovered, turn out to
be that of a relation or friend of some one of
the party engaged in the exhumation. Such
a tale is recorded in the *Monthly Magazine*
for April, 1827 ; there a sailor is pressed into
the service of some students who were anxious
to obtain a body. The subject was safely
brought home, and, on being taken from the
sack, turned out to be the sweetheart of the
sailor, who had just returned from sea, and, not
having heard of his girl's decease, was on his
way to greet her after a long absence from
home. Truth and fiction often agree. There
is a case on record of a child who had died of
scrofula, and whose body was brought to St.
Thomas' Hospital by Holliss, a well-known
resurrectionist. The body was at once recognised by one of the students as that of
his sister's child ; on this being made known
to the authorities at the hospital, the corpse

was immediately buried before any dissection had taken place.

In vols. 1 and 2 of the *Medical Times* there is a series of articles, entitled " The Confessions of Jasper Muddle, Dissecting-room Porter." These papers are signed " Rocket," but were written by Albert Smith.* One of the articles contains an account of a handsome young lady who came to the dissecting-room late at night, and begged for the body of a murderer executed the previous day, which was then being injected, ready for lecture purposes. In the *Tale of Two Cities*, Dickens has given us a good study of a resurrection-man in the person of Mr. Cruncher. Moir in *Mansie Wauch*, Lytton in *Lucretia*, Mrs. Crowe in *Light and Darkness*, and Miss Sergeant in *Dr. Endicott's Experiment*, have also used the body-snatcher in fiction.

As long as the Barber Surgeons kept to their right of the exclusive teaching of anatomy, there was small need of bodies for dissection. This right the Company jealously guarded. On 21st May, 1573, the following entry occurs

* It may be interesting to mention that Albert Smith's remuneration for these papers was five shillings per page of three columns.

in the records, "Here was John Deane and appoynted to brynge in his fyne x^{li} for havinge an Anathomye in his howse contrary to an order in that behalf between this and mydsomer next."* As late as 1714 this rule was put in force against no less a man than William Cheselden. The entry in the books of the Company runs as follows, "At a Court of Assistants of the Company of Barbers and Surgeons, held on the 25th March, 1714. Our Master acquainting the Court that Mr. William Cheselden, a member of this Company, did frequently procure the Dead bodies of Malefactors from the place of execution and dissect the same at his own house, as well during the Company's Publick Lectures as at other times without the leave of the Governors and contrary to the Company's By law in that behalf. By which means it became more difficult for the Beadles to bring away the Companies Bodies and likewise drew away the members of this Company and others from the Public Dissections and Lectures at the Hall. The said Mr. Cheselden was, therefore, called in.

* *Annals of the Barber Surgeons*, by SIDNEY YOUNG, p. 317.

But having submitted himself to the pleasure
of the Court with a promise never to dissect at
the same as the Company had their Lecture at
the Hall, nor without leave of the Governors
for the time being, the said Mr. Cheselden was
excused for what had passed with a reproof for
the same pronounced by the Master at the
desire of the Court."*

By the Act Henry VIII., xxii., cap. 12, provision was made for the Company of Barbers
and Surgeons to have the bodies of malefactors
for the purpose of dissection. This part of the
Act was as follows : "And further be it enacted
by thauctoritie aforesayd, that the sayd maysters
or governours of the mistery and comminaltie
of barbours and surgeons of Londō & their
successours yerely for ever after their sad
discrecions at their free liberte and pleasure
shal and maie have and take without cõtradic-
tion foure persons condempned adjudged and
put to deathe for feloni by the due order of the
Kynges lawe of thys realme for anatomies with
out any further sute or labour to be made to
the kynges highnes his heyres or successors

* SOUTH and D'ARCY POWER, *Memorials of the Craft of
Surgery*, p. 233, *note.*

for the same. And to make incision of the same deade bodies or otherwyse to order the same after their said discrecions at their pleasure for their further and better knowlage instruction in sight learnyng & experience in the sayd scyence or facultie of Surgery."

The "foure bodies" could not always be obtained without difficulty ; despite the precautions of the Company private anatomy was, to a certain extent, carried on, and the bodies of malefactors had a market value. The following entries from the *Annals of the Barber Surgeons* are illustrative of this :

"6th March, 1711.* It is ordered that William Cave, one of the Beadles of this Company, do make Inquiry who the persons were that carryed away the last body from Tyburne, and that such persons be Indicted for the same.

"9th October, 1711. Richard Russell, one of the persons who stands Indicted for carrying away the last publick body applying himself to this Court and offering to be evidence against the rest of the persons concerned It is ordered that the Clerk do apply himself to Her Majesty's

* YOUNG, *loc. cit.* p. 349.

Attorney Generall for a Noli p'sequi as to the said Russell in order to make him an evidence upon the s^d Indictment and particularly ag^st one Samuell Waters whom the Court did likewise order to be indicted for the said fact."

Often there were riots caused by the Beadles of the Company going to Tyburn for the bodies of murderers. This rioting was carried to such an extent that it was found necessary to apply for soldiers to protect the Beadles.

"28th May, 1713. Ordered that the Clerk go to the Secretary at War for a guard in order to gett the next Body [from Tyburn.]"

The dissection of these bodies was made known by public advertisement. The following is from the *Daily Advertiser* of January 15th, 1742: "Notice is hereby given that there being a publick Body at Barbers and Surgeons Hall, the Demonstrations of Anatomy and the Operations of Surgery will be at the Hall this evening and to-morrow at six o'clock precisely in the Amphitheatre."

In 1752 it was ordered that bodies of murderers executed in London and Middlesex should be conveyed to the Hall of the Surgeons Company to be dissected and anatomized, and any attempt to rescue such bodies was made felony.

In 1745 the Barbers and Surgeons, who from 1540, until that date, had formed one Company, separated, and the latter were incorporated under the title of "The Masters, Governors, and Commonalty of the Art and Science of Surgery." To the Surgeons naturally fell the duty of dissecting the bodies of the malefactors handed over for that purpose. The building of the Surgeons' Company was in the Old Bailey; there was, therefore, no difficulty in removing the bodies from Newgate. In 1796 the Company came to a premature end through an improperly constituted Court having been held. It was attempted to put matters right by a Bill in Parliament, but there was so much opposition from those persons who were practising without the diploma of the Corporation, that the Bill, after passing safely through the Commons, was thrown out by the Lords. In the following year attempts were made to come to terms with the opponents of the Bill, and finally it was agreed to petition for a Charter from the Crown to establish a Royal College of Surgeons in London. These negotiations were successfully carried out in 1800, and the old Corporation having disposed of their Old Bailey property to the City Authorities, the College took possession of a house in Lincoln's Inn Fields, the site of part of the present building.

During the debate in the House of Lords on the Bill just mentioned, the Bishop of Bangor, who had charge of the measure, sent for the Clerk of the Company, and informed him that a strong opposition was expected to the Bill, on account of the inconvenience that would arise from the bodies of murderers being conveyed through the streets from Newgate to Lincoln's Inn Fields. To remedy this a clause was proposed, giving the College permission to have a place near to Newgate, where the part of the sentence which related to the dissection of the bodies might be carried out.

That this difficulty of moving the bodies was not a fancied one, the following extract from "Alderman Macaulay's Diary" will show : " Dec. 6, 1796. Francis Dunn and Will. Arnold were yesterday executed for murder and the first malefactors conveyed to the new Surgeons' Hall in the Lincoln's Inn Fields.

They were conveyed in a cart, their heads supported by tea chests for the public to see : I think contrary to all decency and the laws of humanity in a country like this. I hope it will not be repeated."*

Just at this date the Corporation were removing from their old premises to Lincoln's Inn Fields ; the last Court in the Old Bailey was held on October 6th, 1796, and the first at Lincoln's Inn Fields on January 5th, 1797.

In July, 1797, it was reported to the Court that Mr. Chandler, one of their members, "had in the most polite and ready manner offered his stable for the reception of the bodies of the two murderers who were executed last month." The thanks of the Court were voted to Mr. Chandler "for his polite attention to the Company upon that occasion."

After the Bill had been lost in the Lords, the following resolution was passed by the Court in November, 1797 : "Resolved that in order to evince the sincerity of the Court to remove all reasonable objections to the present situation in Lincoln's Inn Fields the Clerk be directed, with proper assistance, to

* *Academy*, vol. vi. p. 208, 1874.

look for a temporary dissecting-room at a place in or near the Old Bailey until a permanent one near the place of execution can be established."

In June, 1800, a warehouse was taken in Castle Street, Cow Cross, West Smithfield, for eighteen months, as, owing to the labours of taking over the Hunterian Collection, there had been no time for obtaining a permanent place. A house in Duke Street, West Smithfield, was afterwards leased for the purpose, and arrangements were made for Pass, the Beadle, to reside there. This landed the College in a small expense, as in 1832 the Beadle was elected Constable of the Ward of Farringdon, and the Council had to pay a fine of £10 in place of his serving the office. At the expiration of the lease of the Duke Street house, so great an increase of rent was demanded that the College gave up the premises, and took a newly-built house in Hosier Lane, on a lease for twenty-one years. Here the dissections were carried on until the passing of the Anatomy Act, when the College had no longer to share with the hangman the duty of carrying out the sentence on murderers who were condemned to be hanged and anatomized.

The bodies were not really dissected by the College Authorities; a sufficient incision was made to satisfy the requirements of the Act, and the body was then handed over to one of the Teachers of Anatomy. The following is a copy of an order authorizing the Secretary of the College to give up a body :

" Ordered.

"That the body of Mary Whittenbach executed this day at the Old Bailey for murder be delivered (after the necessary dissection by the College) to Mr. Joseph Henry Green.

" WILLIAM BLIZARD
" WM. NORRIS
" ANTHY. CARLISLE.

" Royal College of Surgeons
" *17th day of Sept.* 1827
" To Mr. BELFOUR, Secy. to the College."

There is in the Library of the Royal College of Surgeons of England a series of drawings of the heads of murderers, made by the two Clifts, father and son, when the bodies were brought to the College for dissection. These drawings include Bishop and Williams (see p. 107),* and Bellingham, who was executed

* For the portraits of Bishop and Williams see p. 112.

in 1812 for the murder of Mr. Perceval in the lobby of the House of Commons.

Earl Ferrers, who suffered the extreme penalty of the law in 1760 for the murder of his steward, was taken to Surgeons' Hall, where an incision was made in the body ; instead of being further dissected it was given over to the relatives for burial.

At the execution of Bishop and Williams the Sheriffs of London felt that some means should be taken to show gratitude to Mr. Partridge, and the other officials of King's College, for the way they had brought the murderers to justice. The following letter was therefore addressed to the College of Surgeons :

" JUSTICE HALL,
"*Dec.* 5, 1831.

" *To the Governors and Directors of the College of Surgeons.*

" It is our particular desire and we do ask that it may be thought but a reasonable request that the bodies of the malefactors executed in the front of Newgate this morning should be sent to King's College—by the vigilance of whose surgical establishment these offenders were detected and ultimately brought to justice, we shall therefore feel obliged by

your handing over these bodies to the King's
College.

> " We are, with great respect,
> " Your mo. ob. Servts.,
>
>   " J. Cowan } *Sheriffs.*"
>   " John Pirie }

The body of Bishop was given to Mr.
Partridge, and that of Williams went to Mr.
Guthrie at the Little Windmill Street School
of Anatomy.

The following account of the reception of
one of the bodies is by Mr. T. Madden Stone,
for many years an official at the College. It
was printed in a series of articles, entitled
" Echoes from the College of Surgeons."*

" The executions generally took place at
eight o'clock on Mondays, and the ' cut down,'
as it is called, at nine, although there was no
cutting at all, as the rope, with a large knot
at the end, was simply passed through a thick
and strong ring, with a screw, which firmly
held the rope in its place, and when all was
over, Calcraft, *alias* ' Jack Ketch,' would make
his appearance on the scaffold, and by simply
turning the screw, the body would fall down.

---

* *Hospital Gazette*, from Sep. 13, 1890, to March 7, 1891.

At once it would be placed in one of those large carts with collapsible sides, only to be seen in the neighbourhood of the Docks, and then preceded by the City Marshal in his cocked hat, and, in fact, all his war paint, with Calcraft and his assistant in the cart, the procession would make its way to 33 Hosier Lane, West Smithfield, in the front drawing room of which were assembled Sir William Blizard, President of the Royal College of Surgeons, and members of the Court desirous of being present, with Messrs. Clift (senior and junior), Belfour, and myself. On extraordinary occasions visitors were admitted by special favour. The bodies would then be stripped, and the clothes removed by Calcraft as his valuable perquisites, which, with the fatal rope, were afterwards exhibited to the morbidly curious, at so much per head, at some favoured public-house. It was the duty of the City Marshal to be present to see the body 'anatomised,' as the Act of Parliament had it. A crucial incision in the chest was enough to satisfy the important City functionary above referred to, and he would soon beat a hasty retreat, on his gaily-decked charger, to report the due execution of his duty. These experiments concluded, the body would be stitched up, and Pearson, an old museum attendant, would remove it in a light cart to the hospital, to which it was intended to present it for dissection."

These bodies of murderers were the only ones which could be legally used for dissection; it is therefore obvious that the number was quite insufficient for the wants of the Metropolitan Schools, and the teachers were thus forced to obtain a supply from other sources.

It was strongly urged, but urged in vain, that the whole difficulty would disappear if a short Act were passed, doing away with the dissection of murderers, and enacting that the bodies of all unknown persons who died in workhouses or hospitals, without friends, should be handed over, under proper control, to the different teachers of anatomy. That these would be sufficient was afterwards made clear by the Committee on Anatomy.*   In their Report

* This Committee was appointed by the House of Commons in 1828, to take evidence and report on the necessity of obtaining bodies for anatomical purposes. The work of the Committee is referred to at greater length on p. 102.

it is stated that the returns obtained from 127 of the parishes situate in London, Westminster, and Southwark, or their immediate vicinity, showed that out of 3744 persons who died in the workhouses of these parishes in the year 1827, 3103 were buried at the parish expense, and that of these about 1108 were not attended to their graves by any relations. The number of bodies obtained from this source would have exceeded those supplied by the resurrection-men, and would have been adequate for the wants of the London Schools.

The newspapers of the day contain many proposed solutions of the difficulty. One correspondent gravely suggested that as prostitutes had, by their bodies during life, been engaged in corrupting mankind, it was only right that after death those bodies should be handed over to be dissected for the public good. Another correspondent proposed that all bodies of suicides should be used for dissection, and that all those persons who came to their death by duelling, prize-fighting, or drunkenness, should be handed over to the surgeons for a similar purpose.

Mr. Dermott, the proprietor of the Gerrard

c

Street, or Little Windmill Street, School of
Medicine, proposed a scheme by which a fund
was to be raised by grants from Government,
and from the College of Surgeons, and by
voluntary contributions from the nobility and
gentry.   This fund was to be invested in the
names of "opulent and respectable men," not
more than one-third of whom were to be
members of the medical profession.   It was
proposed to expend the interest on this fund
in paying a sum not exceeding seven pounds
to those persons who were willing to contract
for the sale of their bodies for dissection.
Registers were to be kept of all such persons,
and the Committee were to have the power
of claiming the body six hours after death.
Mr. Dermott also suggested that all medical
men should leave their own bodies to be used
for anatomical teaching.   It is hardly necessary
to point out the absurdity of the first part
of this scheme ; the Committee, after paying
their seven pounds, would have had no control
over the subsequent movements of the persons
whose bodies they had thus purchased, and
it was hardly to be expected that friends of the
deceased would send notice to the Committee

that the body was ready for them.   Both parts
of the scheme would have required an Act
of Parliament, as executors were not bound
to give up a corpse, even though instructions
had been left that it was a person's wish
that his body should be used for anatomical
purposes.   Many such bequests have been
made, and in some instances the desire of
the testator has been carried into effect.   To
try to do away with some of the prejudices
against dissection, Jeremy Bentham left his
body for this purpose ; the dissection was duly
carried out at the Webb Street School, and
at the request of Dr. Southwood Smith, Mr.
Grainger delivered the following oration over
the body on June 9th, 1832 :

"Gentlemen,—In presenting myself before
you this day, at the request of my friend and
colleague, Dr. Southwood Smith, I can assure
you I do so strongly impressed with the high
importance of the duty I have undertaken, and
the responsibility I have thus assumed.   Gentlemen, it is no ordinary occasion on which we are
assembled.   We are here collected to carry
into execution the last wishes of one whose
mortal career, prolonged far beyond the usual

limits of man's existence, has been devoted with
almost unexampled energy and perseverance
to the establishment of those great moral and
political truths, on which the happiness and the
enlightenment of the human race are founded.
Ill would it become me, however, to dwell on
the genius, the philanthropy, or the integrity
of the illustrious deceased.   His eulogium has
already been eloquently pronounced by one
more fitted to do justice to such an under-
taking than the humble individual who now
addresses you.   It would be more suitable
to the object of the present meeting that I
should consider in what manner the intentions
of the late Mr. Bentham, regarding the disposi-
tion of his remains, can best be carried into
effect.   But before I do this, it may be proper
to inform some of my auditors what those
intentions were.   This great man was an
ardent admirer of the science of medicine,
and his penetrating mind was not slow in
perceiving that the safe and successful practice
of the healing art entirely rests on a thorough
knowledge of the natural structure and functions of the human body.   He also perceived
that there was but one method of obtaining

such knowledge, viz., dissection. In proceeding to inquire how it came to happen that in a country like England, justly proud of those numerous institutions in which science is so successfully cultivated, so little encouragement, or more correctly speaking, so much opposition, was offered to the advancement of so indispensable a branch of knowledge, Mr. Bentham discovered that this repugnance to dissection sprang from a feeling strongly implanted in the human breast—a feeling of reverence towards the dead. Far be it from me to condemn such a sentiment, for it has its source in some of the purest principles of our nature. But if it can be shown that an undue indulgence in this feeling produces incalculable mischief in society, it becomes the duty of all who are interested in the happiness of mankind to oppose the progress of such injurious opinions. Mr. Bentham, impressed with this idea, and thinking it unjust that the humbler classes of the community should alone be called upon to sacrifice those feelings which are cherished alike by the rich and poor, determined to devote his own body to the public good. He knew that this determination would inflict pain on many of his dearest friends. An example of this character, emanating from a person so talented, so influential, and so esteemed, is calculated to operate a most beneficial effect on the public mind, and I cannot refrain from considering the dissection of the body now before us as an important era in the progress of anatomy, as it is one of the first that in this country has been employed for the purposes of science, under the direct sanction of the individual expressed during his lifetime ; he also knew that obstacles would probably be offered to its fulfilment, but with an indifference to personal feeling rarely witnessed, he took effectual means to carry his resolution into effect. And thus, gentlemen, did the last act of this illustrious man's ex- istence accord with that leading principle of his well-spent life—the desire to promote the universal happiness and welfare of mankind."

Bentham's skeleton, clothed in his usual attire, is now in University College, London.

Messenger Monsey, the eccentric physician to the Chelsea Hospital, was exceedingly anxious that his body should be examined after death. He obtained a promise from

Mr. Forster, of Union Court, that he would perform this service for him. So anxious was Monsey for the *post mortem* to be carried out, that in May, 1787, he wrote to Cruikshank, the anatomist, as follows :

"Mr. Foster (*sic*) a Surgeon in Union Court, Broad Street, has been so good as to promise to open my Carcass and see what is the matter with my Heart, Arteries, Kidnies, &c. He is gone to Norwich and may not return before I am [dead]. Will you be so good as to let me send it to you, or if he comes will you like to be present at the dissection. I am now very ill and hardly see to scrawl this & feel as if I should live two days, the sooner the better. I am, tho' unknown to you

"Your respectfull humble Servant
"MESSR. MONSEY."

Monsey lived until December 20th, 1788 ; his wishes were duly carried out by Mr. Forster, at Guy's Hospital, in the presence of the students.

Ninety-nine gentlemen of Dublin signed a document, in which the wish was expressed that their bodies, instead of being interred, should be devoted by their surviving friends "to the more rational, benevolent, and honourable purpose of explaining the structure, functions and diseases of the human being."

A Mr. Boys, who died in 1835, wished to be made into "essential salts" for the use of his female friends. In a letter to Dr. Campbell, written four years before his death, he asks : "Are you now disposed (without Burking) to accomplish my wish, when my breath or spirit shall have ceased to animate my carcase, to perform the operation of vitrifying my bones, and sublimating the rest, thereby cheating the Devil of his due, according to the ideas of some devotees among Christians? And, that I may not offend the delicate olfactory nerves of my female friends with a mass of putridity, if it be possible, let me rather fill a few little bottles of essential salts therefrom, and revive their drooping spirits. It may be irksome to you to superintend the business, but, perhaps, you have knowledge of some rising genius or geniuses who may be glad of a subject without paying for it. Let them slash and cut, and divide, as best please 'em."

The following account, taken from a newspaper of 1810, shows that untoward events sometimes followed a request of this kind. A journeyman tailor died at the *Black Prince*, in Chandos Street, and directed, in his will, that his body should be opened in the presence of Mr. Wood, the landlord. This instruction was carried out. The paragraph goes on to say that the dissection was scarcely concluded "when the landlord, a stranger to such exhibitions, was seized with sickness and vomiting ; and, on reaching the bar, was prevailed upon by his wife to take a glass of brandy and water ; in a few minutes he was obliged to be carried to bed, never to rise again ; on Friday last, the third day from the attack, he died in a state of delirium, not from contagion, or a predisposition to disease, but solely from the impression made upon his mind by the anatomical performance, which, he observed, exceeded in horror any thing he had ever beheld."

It was not an uncommon thing for persons to try to put into effect part of Dermott's plan, by offering to leave their bodies for anatomical purposes, on the condition that they were paid a certain sum down. This was generally only a swindling dodge, and one by which the teachers were not to be caught, as they could have no hold on the persons whose bodies they purchased, nor could they compel the friends to give them up after death. The following letter, preserved amongst Sir Astley Cooper's papers, and now forming part of the Stone Collection at the Royal College of Surgeons of England, is a specimen :

" SIR,—I have been informed you are in the habit of purchasing bodys and allowing the person a sum weekly ; knowing a poor woman that is desirous of doing so, I have taken the liberty of calling to know the truth.

"I remain, your humble servant."

*

On the back Sir Astley has written, " The *truth* is that you deserve to be hanged for such an unfeeling offer.   A. C."

The idea at the present day has not died out ; quite recently a man called at the College of Surgeons, and offered to sell his body for a cash payment.   It is a fairly common experience of Curators of Pathological Museums to have similar offers from persons suffering from a rare disease, or a curious deformity.

* The letter has no signature.

MORTSAFE IN GREYFRIARS CHURCHYARD, EDINBURGH.

CHAPTER II.

AS has been stated in the previous chapter, there was no need of the resurrection-men, so long as the teaching of anatomy was confined to the Company of Barbers and Surgeons. It has also been pointed out that, as late as 1714, Cheselden was reprimanded for having anatomical demonstrations at his private house. Soon after this date, however, began the establishment of private schools. Mr. Nourse, of St. Bartholomew's, was one of the first to deliver public lectures at his own house. After a time this probably became inconvenient, as we find his advertisement, in 1739, worded thus :

"ANATOMY.

"Designing to have no more lectures at my own house, I think it proper to advertise that I shall begin a Course of Anatomy, Chirurgical Operations and Bandages on Monday, the 11th of Nov., at St. Bartholomew's Hospital.

"EDW. NOURSE, Assistant Surgeon
and Lithotomist to the said Hospital."

Percivall Pott, who was apprenticed to Nourse, followed his master's example, and lectured on Surgery. In 1737 we find Dr. Fr. Nicholls advertising thus :

" On Wednesday, the 2nd of February, at the House below the Bull Head, in Lincoln's Inn Fields, at five in the evening, will begin a Course of Anatomy and Physiology, introductory to the study and practice of Physick in all its branches by Fr. Nicholls, M.D. N.B. A compendium referring to the several matters, explain'd in these Lectures, is sold by John Clarke, under the Royal Exchange, and F. Woodward, at the Half Moon, within Temple Bar, Booksellers."

The following is the advertisement of Cæsar Hawkins, from a newspaper of 1739 :

" In Pall Mall Court, in Pall Mall. On Thursday, the 5th of February next, will begin a Course of Anatomy, with the principal Operations in Surgery and their suitable Bandages, by Cæsar Hawkins, Surgeon to St. George's Hospital."

Joshua Brookes' advertisement, in 1814, ran as follows :

" THEATRE OF ANATOMY, BLENHEIM STREET,
GREAT MARLBOROUGH STREET.

" The Summer Course of Lectures on Anatomy, Physiology, and Surgery, will be commenced on Monday, the 6th of June, at seven o'clock in the

morning. By Mr. Brookes.—Anatomical Converza-
tiones will be held weekly, when the different
Subjects treated of will be discussed familiarly, and
the Students' views forwarded. To these none but
Pupils can be admitted. Spacious Apartments,
thoroughly ventilated, and replete with every
convenience, will be open at five o'clock in the
morning, for the purposes of Dissecting and Injecting,
when Mr. Brookes attends to direct the Students
and demonstrate the various parts as they appear
on Dissection.

" The inconveniences usually attending Anatomical
Investigations, are counteracted by an antiseptic
process. Pupils may be accommodated in the
House. Gentlemen established in Practice, desirous
of renewing their Anatomical Knowledge, may be
accommodated with an apartment to dissect in
privately."

A very interesting account of the old
Anatomical Schools, by Mr. D'Arcy Power,
will be found in the *British Medical Journal*,
1895, vol. 2, p. 141. The paper is entitled
"The Rise and Fall of the Private Medical
Schools in London." It has been reprinted,
with other articles, in a pamphlet, entitled
The Medical Institutions of London.

In Great Britain, as no licence was required
for opening an Anatomical School, there was
no limit to their number; there was also

no regular legal supply of subjects, except
the bodies of murderers, executed in London
and the county of Middlesex, which came to
the schools through the College of Surgeons.
In Paris a licence had to be obtained before
opening an Anatomical School, and bodies
were regularly supplied to the licensed
places.

With the rise and competition of the
Medical Schools in London, the difficulty
of getting an adequate number of bodies
increased. The absolute necessity of having
a good supply for the use of students,
so as to prevent them from going off to rival
schools, caused the teachers to offer large
prices, and thus made it worth while for men
to devote themselves entirely to obtaining
bodies for this purpose. At first the trade
was carried on by a very few men, and without
any public scandal, but the inducements
mentioned above enticed others into the
business ; these were of the lowest class, often
professed thieves, and the fights and disputes
of these men, one with the other, in church-
yards, often made really more scandal than the
actual stealing of the bodies. It was stated

by the police in 1828 that the number of persons who, in London, lived regularly on the profits of exhumation, did not exceed ten ; but there were, in addition to these, about two hundred who were occasionally employed. These latter individuals were thieves of the lowest grade, and the most desperate and abandoned class of the community. The men worked generally in gangs, and would do anything to spoil the success of their opponents in the business. If a body were bought by one of the teachers from an outside source, the regular men would sometimes break into the dissecting-room and cut the body in such a manner as to make it useless for anatomical purposes. If this could not be done, they would give information to the police that a stolen body was lying in a certain dissecting-room. Joshua Brookes, the proprietor of the Blenheim Street, or Great Marlborough Street, School, was a victim in this way ; a body, for which he had paid 16 guineas, was taken away from his school through information of this kind, and the police officer who carried out the business was, as a reward for his efforts, presented with

D

a silver staff, purchased by public subscription. Brookes seems to have got on very badly with the resurrection-men ; at one time, because he refused five guineas as a douceur at the beginning of the session, two dead bodies, in a high state of decomposition, were dropped at night close to his school by the men whom he had thus offended ; one of these bodies was placed at the Poland Street end of Great Marlborough Street, and the other at the end of Blenheim Street. Two young ladies stumbled over one of these bodies, and at once raised such a commotion that, had it not been for the prompt assistance of Sir Robert Baker and the police, Brookes would have fared very badly at the hands of the mob which soon collected. The fact of his house being near to the Marlborough Police Court, on more than one occasion saved Brookes from the popular fury.

A subject was brought to him one day in a sack, and paid for at once ; soon after it was discovered that the occupant of the sack was alive. This was not a case of attempted murder ; the "subject" was a confederate of those from whom he had been purchased, and had, in all probability, been thus introduced to the premises for purposes of burglary.

The competition of the schools had risen to such a height in the demand for bodies, that Brookes stated that for a subject, which would have cost two guineas in his student days, he had paid as much as sixteen guineas. Nor was the cost of the body the only expense to the teacher. At the beginning of each session he was waited upon by the resurrection-men, who offered to supply him regularly with bodies at a fixed price, on the condition that a douceur was paid down at once. The teachers were powerless in the matter, and had either to accede to the offered terms, or to lose their students through not having a sufficient supply of subjects. The scarcity of bodies was most keenly felt at the beginning of the session ; the resurrection-men knew that they could command their own terms, and would not supply any subjects until the teachers had conceded all their demands. This was felt to be bad for the students, and Dr. James Somerville, who was assistant to Brodie at the Great Windmill Street School, in giving evidence before the Committee on Anatomy, said that " the pupils not being able to proceed for a certain time lose their ardour, and get into habits of idleness."

At the end of the session the resurrection-men again waited on the proprietors of the schools, and demanded "finishing money." In some papers relating to Sir Astley Cooper, which were referred to in a letter published in the *Medical Times*, 1883, vol. 1, p. 343, we read : " May 10th, 1827, Paid Hollis, Vaughan, and Llewellyn, finishing money, £6 6s. od. 1829, June 18th, Paid Murphy, Wildes, & Naples, finishing money £6 6s. od."

The cost of the bodies in this way to the teachers was more than they could charge to the students, and the deficiency thus created was made up by increased fees for the lectures. The expenses, moreover, did not end here. If one of the resurrection-men was unfortunate enough to get a term of imprisonment, the teacher had to partly keep the man's wife and family whilst he was serving his sentence. A solatium was also expected on his release from gaol. Mr. R. D. Grainger spent £50 in this way for one man, and several guineas in keeping the family of another Resurrectionist

whilst the latter was in gaol. Sir Astley
Cooper is known to have spent large sums
of money for a similar purpose. The following may be cited as examples : " January 29th,
1828, Paid Mr. Cock to pay Mr. South half
the expenses of bailing Vaughan from Yarmouth and going down £14 7s. od. 1829,
May 6th, Paid Vaughan's wife 6s. Paid
Vaughan for twenty-six weeks' confinement
at 10s. per week, £13 os. od."

If any independence were shown by the
teachers, and the demands of the men resisted,
victory generally fell to the lot of the Resurrectionists. A teacher, perhaps, would refuse to
pay the exorbitant demands, and would employ
other men to obtain bodies for him. These
were then watched by the regular gang, and
information to the police was laid against them
on every occasion. The bodies obtained by
the irregular men were often taken from them
by those who considered they had a monopoly
in the business ; these subjects were then
hacked and cut about so as to make them quite
useless for anatomical purposes. So the
supply at this particular school would be
very short, and great indignation would arise

amongst the students, who had paid their fees, and therefore demanded an adequate number of bodies for dissection. The teacher was thus obliged to give way, and to accede to the demands of the regular gang.

The teachers formed themselves into an Anatomical Club for their own protection ; by this means it was hoped to regulate the price to be paid for bodies, by agreement amongst the members of the Club not to give more than a certain amount. This agreement does not seem, according to Mr. South, to have been very faithfully kept, and so, with new schools springing up and giving rise to still greater competition, the teachers were as much as ever in the hands of the resurrection-men.

It must not be supposed that all the bodies which were supplied to the schools were exhumed. Many of them were stolen or obtained by false pretences before burial. Glennon, the police officer, who has been before mentioned in connection with Joshua Brookes, told the Committee that he had recovered between fifty and a hundred bodies for persons who had had their houses broken open, and bodies stolen from them whilst in the coffin awaiting burial. The following case, tried at the London Sessions in 1830, is an example of this :

"LONDON ADJOURNED SESSIONS.

" TUESDAY.—BODY-SNATCHING. — A well-known pilferer of graves, named Clarke, was tried upon an indictment, charging him with having stolen the body of a dead child, aged about four years, which had been under the care of a nurse named Mary Hopkins. The facts which came out in evidence are as follows : The deceased was the daughter of a woman of the town, residing in Shire Lane, and had been kept at the nurse's lodging, which was in the same neighbourhood. She died on a Friday, and Clarke, whose ears were described as 'quick to the toll of the passing bell,' paid the nurse a visit the next morning, under pretence of hiring a cellar under the house. He took occasion to notice the poor woman's son ; said it was a pity to see the boy idle, and that he should have immediate employment, and called again with evidences of still stronger interest in favour of the family. ' By the way,' said he, ' I understand you have had a death

lately.' 'Yes, sir,' said the nurse, 'a poor little
girl is departed.' 'Poor little dear,' cried the
snatcher, 'I should like to look at the little
innocent.' He was forthwith led into the front
parlour, where the body lay in a coffin, and
observing that its position was favourable to
his intention, he sympathized with the nurse,
and said, 'We must all come to this sooner
or later,' and then he went to get a half-pint of
summut to comfort them. The nurse disposed
of a glass, which presently set her in a
profound sleep, and when she awoke the body
of the babe was gone. It appeared that the
snatcher, after having quitted the house, as
if for good, returned, and opening the parlour-
window hooked out with a stick the corpse
of the child, and went off with it towards a
market that is open at all hours, near Bridge-
water Square. However, a police officer, who
knew his trade, laid hands upon him, telling
him he was wanted. The snatcher then threw
down the child and took to his heels, but was
apprehended and lodged in the Compter.
The nurse proved the identity of the body.
Upon her cross-examination, by Mr. Payne,
she stated that the mother had not been to see

the deceased for four or five days before the death. The Jury returned a verdict of Guilty, but some of them audibly spoke of recommending the prisoner to mercy, but made no appendage to that effect. The Recorder sentenced the prisoner to be imprisoned for the space of six calendar months."

Sometimes these stolen bodies were claimed after payment had been made to the resurrection-men, but before any dissection had taken place. The following refers to Guy's Hospital : " Returned to Vestry Clerk of Newington, by order of the Treasurer, one male and two females, purchased of Page, &c., on the 25th, who had broken open the dead-house to obtain them."

Bodies of suicides, and of those who had met with an accidental death, were frequently stolen whilst they were awaiting the coroner's inquest. Often in these cases the body-thieves, after selling the subject to a teacher of anatomy, secretly gave information to the police where the missing body might be found. It was then seized by the police, and, after the inquest, handed over to those who claimed to be relatives ; these supposed relatives were frequently confederates of the thieves, and by them the body was at once taken off and again sold to another teacher.

The following case is from a newspaper of 1823:

" SUICIDE AND THE BODY STOLEN.—Tuesday evening last a young woman of respectable and interesting appearance was observed for some time parading the banks of the Surrey Canal, Camberwell, in a melancholy mood, and at length she plunged into the water; on which a man rushed in after her and dived several times, but failed in recovering the body, which was not found till the following morning, when it was taken to the Albany Arms, near the Canal, for the Coroner's inquest, which was to have taken place on Thursday. On the landlord proceeding to the shed on Wednesday morning, where the body had been deposited, he discovered, that in the course of the night, it had been broken open, and the corpse of the female stolen away. He instantly repaired to the Police Office, Union Street, and gave information of the circumstance to the Magistrates, who gave orders that immediate inquiry should be made at Mr. Brookes's, where the body has since been discovered and given up. The poor woman was unclaimed, and the verdict of the Coroner's Jury was ' Found Drowned.' "

A favourite trick, in the carrying out of which a woman was generally necessary, was that of claiming the bodies of friendless persons who died in workhouses, or similar institutions. Immediately it was found out that such an one was dead a man and woman, decently clad in mourning, in great grief, and often in tears, called at the workhouse to take away the body of their dear departed relative. If the trick proved successful, as it often did, the body was taken straight off to one of the schools and sold. The parish authorities, probably, were not over particular about giving up the body, if the deceased were a stranger, as by this means they saved the cost of burial.

Subjects, too, were obtained from cheap undertakers, who kept the bodies of the poor until the time for burial. The coffin was weighted so as to conceal the fraud, and the mockery of reading the Burial Service over it was gone through in the presence of the unsuspecting relatives.

That some bodies were obtained by murder there can be no doubt. The exposure caused by the trials of Burke and Hare in Edinburgh, and Bishop and Williams in London, proves this.

The facts previously stated, however, go very far to exonerate the anatomists from the false charge (freely made at the time) of their being privy to these murders. It has been frequently stated that signs of murder could be easily seen, and that the fact of the body being fresh, and there being no evidence of its having been interred, ought to have at once suggested foul play, and to have caused the teacher to communicate with the police. But it must be remembered that the murders were generally very artfully contrived by suffocation, so as to leave no outward signs of ill-treatment. It was also no uncommon thing, for the reasons just given, to receive at the schools bodies in quite a fresh state, which had evidently never received sepulture.

An account of the *post mortem* on the Italian boy, for whose murder Bishop and Williams were hanged,* has been preserved by Mr.

* See also p. 107.

Clarke.* The examination of the body was carried out by Mr. Wetherfield, of Southampton Street. There were also present Mr. Mayo, Lecturer on Anatomy at King's College; Mr. Partridge, his demonstrator; Mr. Beaman, Parish Surgeon; and his Assistant, Mr. D. Edwards, and Mr. Clarke. The boy's teeth had been removed and sold to a dentist, but beyond this there were no external marks of violence on any part of the body. The internal organs were carefully examined, but no trace of injury or poison could be found. Mr. Mayo, who had a peculiar way of standing very upright with his hands in his breeches' pockets, said, with a kind of lisp he had, " By Jove! the boy died a nathral death." Mr. Partridge and Mr. Beaman, however, suggested that the spine had not been examined, and after a consultation it was decided to do this. It was then found that one or more of the upper cervical vertebræ were fractured. " By Jove!" said Mr. Mayo, " this boy was murthered." The conviction of Bishop and Williams was due, in a very great measure, to Mr. Partridge and Mr. Beaman.

* *Autobiographical Recollections of the Medical Profession,* p. 101.

At the present day it is well-nigh impossible to understand the relations between men of honour and education, such as the teachers of anatomy were, and the ruffians who carried on this ghastly trade. It must, however, be borne in mind that, until the passing of the Anatomy Act in 1832, there was no provision for supplying the means by which the student might be taught this necessary part of his professional education ; the only way in which teachers could get material for giving instruction was by dealing with the resurrection-men.

It would have been quite impossible for the resurrection-men to have obtained the number of bodies they frequently did, had they not been able to bribe the custodians of the different burial-grounds. Sometimes they met with a difficulty in the shape of a keeper newly appointed to replace one who had been dismissed for being privy to these depredations. In most instances this was soon overcome ; if, at the outset, the custodian could not be bribed, he could generally be induced to drink, and then, whilst he was in a state of intoxication, the body which the resurrection-men wished to obtain could be easily removed.

After this first step there was generally very little difficulty in the future.

Sometimes, too, the grave-diggers not only gave information to the Resurrectionists, but acted as principals themselves. In Benson's *Remarkable Trials* is recorded the case of John Holmes, Peter Williams, and Esther Donaldson. Holmes was grave-digger at St. George's, Bloomsbury; Williams was his assistant, and Donaldson was charged as an accomplice. They were prosecuted before Sir John Hawkins at the Guildhall, Westminster, in December, 1777, for stealing the body of Mrs. Jane Sainsbury, who died in the previous October, and was buried in the St. George's burial-ground. Holmes and Williams were sentenced to six months' imprisonment, and to be whipped on their bare backs from the end of Kingsgate Street, Holborn, to Dyot Street, St. Giles. The sentence, says Benson, was duly carried out amidst crowds of well-satisfied and approving spectators. The woman Donaldson was acquitted.

The ranks of the resurrection-men were largely recruited from the keepers of burial-grounds. When these men had lost their

situations for connivance at the stealing of
bodies, they naturally joined their old associates,
and became part of the regular gang.

The bribery of the custodians will account
for the large number of bodies often obtained
in one night. Had there been the slightest
vigilance on the part of the authorities, it
would have been absolutely impossible for the
resurrection-men to have spent the time
necessary for their work without detection.
The amount of time required for the work
depended greatly on the soil. One man told
Bransby Cooper that he had taken two bodies
from separate graves of considerable depth,
and had restored the coffins and the earth to
their former positions in an hour and a half.
Another man said that he had completed the
exhumation of a body in a quarter of an hour ;
but in this instance the grave was extremely
shallow, and the earth loose and without
stones. If much gravel had to be dug
through, the resurrection-men had a peculiar
way of using their spades, so that the gravel
was thrown out of the grave quite noiselessly.

On Thursday, February 20th, 1812, the
Diary tells us that 15 large bodies and one

small one were obtained from St. Pancras.
No doubt this was simplified by the custom
of burying several paupers in one grave. To
obtain these it was necessary to dig all the
earth out, so that each coffin could be dealt
with; the men generally worked very soon
after a funeral, and so the earth was much more
easily moved than it would have been if they
had been obliged to dig through undisturbed
ground. When only one body was to be had,
a small opening was dug down to the head
of the coffin, which was then broken open, and
the body was pulled up with a rope, fastened
either round the neck or under the armpits.

In a memoir of Thomas Wakley, the
founder of *The Lancet*,* the following account
of the *modus operandi* of the resurrection-men
is given : " In the case of a neat, or not quite
new grave, the ingenuity of the Resurrectionist
came into play. Several feet—fifteen or
twenty—away from the head or foot of the
grave, he would remove a square of turf,
about eighteen or twenty inches in diameter.
This he would carefully put by, and then
commence to mine. Most pauper graves were

* *Lancet*, 1896, vol. i, p. 187.

of the same depth, and, if the sepulchre was that of a person of importance, the depth of the grave could be pretty well estimated by the nature of the soil thrown up. Taking a five-foot grave, the coffin lid would be about four feet from the surface. A rough slanting tunnel, some five yards long, would, therefore, have to be constructed, so as to impinge exactly on the coffin head. This being at last struck (no very simple task), the coffin was lugged up by hooks to the surface, or, prefer-ably, the end of the coffin was wrenched off with hooks while still in the shelter of the tunnel, and the scalp or feet of the corpse secured through the open end, and the body pulled out, leaving the coffin almost intact and unmoved.

"The body once obtained, the narrow shaft was easily filled up and the sod of turf accurately replaced.   The friends of the deceased, seeing that the earth *over* his grave was not disturbed, would flatter themselves that the body had escaped the Resurrectionist ; but they seldom noticed the neatly-placed square of turf, some feet away."

A somewhat similar account is given in

the *Memorials of John Flint South.** This
method is also referred to by Bransby
Cooper,† who states that it was told him by
one "who fancied he had found out their
secret, but had, no doubt, been deceived by
some of them purposely." Bransby Cooper
also says that he asked one of the principal
resurrection-men as to the feasibility of this
method, and the man showed him several
objections to it, and stated that "it would
never do." This statement was made after
the resurrection-days were over, when there
could be no advantage in keeping the true
plan secret. It must be remembered that
there were some amateur body-snatchers, and
that it was not at all unlikely that the regular
men would tell to them a plan as full of
difficulties as that quoted above. To make the
tunnel as described, would be impossible, and
it is somewhat difficult to see how grappling-
irons were fastened to the coffin ; a man could
hardly get down a tunnel 18 in. in diameter
and 15 feet in length to do this ; if he did
succeed, his difficulties in returning must have

---

* *Memorials of John Flint South,* by C. T. FELTOE, 1884,
p. 100.   † *Life of Sir Astley Cooper,* vol. i. p. 354.

been still greater. To pull a body out of the
head or foot of a coffin, as described, is an
impossibility. No allowance is made, either, in
digging the tunnel for obstacles, in the shape of
intervening graves or grave-stones. As regards
the evidence on the surface of a grave having
been disturbed, it would be greater in one
opened in this manner than if the recently-
disturbed earth had been again dug out. It
would be impossible to get back into the
tunnel all the earth dug out in the course of
its construction, and this loose earth would
at once attract attention. Generally, bodies
were removed before the graves were finally
tidied up, so that it was difficult to notice a
fresh disturbance.

The writer of the Diary was a cemetery-
keeper when he first began his resurrection
proceedings ; his *modus operandi*, in some
cases, was to take the body out of the coffin,
and place it in a sack, before he began to fill
in the grave. Then, as he gradually threw
the earth in, he kept pulling the sack to the
surface, so that when his work of filling in was
completed, he had the sack close to the top of
the grave. He had then only to wait until

night, when he was able, under cover of the darkness, to remove the body without fear of detection. When the resurrection-men had been successful in their night's work, they were glad to find a temporary shelter for the bodies, as near at hand as possible. This was generally an out-house belonging to one of the schools which they regularly supplied ; the men were permitted to place the bodies there for the night, and to fetch them away the next day. This explains some of the entries in the Diary, such as " Took the whole to ——," and the next day, " Removed the whole from ——." Before removing any of the bodies, the men would find out exactly where they were wanted, and so would save much risk of being arrested with the bodies in their possession.

If the following broadside could be believed, the resurrection-men sometimes performed a valuable service to those who had been buried—

"MIRACULOUS CIRCUMSTANCE :
" *Being a full and particular account of John Macintire, who was buried alive, in Edinburgh, on the 15th day of April, 1824, while in a*

trance, and who was taken up by the resur-
rection-men, and sold to the doctors to be
dissected, with a full account of the many
strange and wonderful things which he saw
and felt while he was in that state, the whole
being taken from his own words.

"I had been some time ill of a low and
lingering fever. My strength gradually wasted,
and I could see by the doctor that I had
nothing to hope. One day, towards evening,
I was seized with strange and indescribable
quiverings. I saw around my bed, innumer-
able strange faces; they were bright and
visionary, and without bodies. There was
light and solemnity, and I tried to move, but
could not; I could recollect, with perfectness,
but the power of motion had departed. I
heard the sound of weeping at my pillow, and
the voice of the nurse say, 'He is dead.' I
cannot describe what I felt at these words. I
exerted my utmost power to stir myself, but
I could not move even an eyelid. My father
drew his hand over my face and closed my
eyelids. The world was then darkened, but I
could still hear, and feel and suffer. For
three days a number of friends called to see

me. I heard them in low accents speak of
what I was, and more than one touched me
with his finger. The coffin was then procured,
and I was laid in it. I felt the coffin lifted
and borne away. I heard and felt it placed in
the hearse; it halted, and the coffin was taken
out. I felt myself carried on the shoulders of
men; I heard the cords of the coffin moved.
I felt it swing as dependent by them. It was
lowered and rested upon the bottom of the
grave. Dreadful was the effort I then made
to exert the power of action, but my whole
frame was immovable. The sound of the
rattling mould as it covered me, was far more
tremendous than thunder. This also ceased,
and all was silent. This is death, thought I,
and soon the worms will be crawling about my
flesh. In the contemplation of this hideous
thought, I heard a low sound in the earth over
me, and I fancied that the worms and reptiles
were coming. The sound continued to grow
louder and nearer. Can it be possible, thought
I, that my friends suspect that they have
buried me too soon? The hope was truly like
bursting through the gloom of death. The
sound ceased. They dragged me out of the

coffin by the head, and carried me swiftly away. When borne to some distance, I was thrown down like a clod, and by the inter-change of one or two brief sentences, I dis-covered that I was in the hands of two of those robbers, who live by plundering the grave, and selling the bodies of parents, and children, and friends. Being rudely stripped of my shroud, I was placed naked on a table. In a short time I heard by the bustle in the room that the doctors and students were assembling. When all was ready the Demon-strator took his knife, and pierced my bosom. I felt a dreadful crackling, as it were, throughout my whole frame; a convulsive shudder instantly followed, and a shriek of horror rose from all present. The ice of death was broken up; my trance was ended. The utmost exertions were made to restore me, and in the course of an hour I was in full possession of all my faculties.

"STEPHENSON, PRINTER, GATESHEAD."

It was quite necessary for the Committee on Anatomy to adopt some means to protect the resurrection-men who gave evidence before it ;

this was done by suppressing their names, and using letters of the alphabet to distinguish the witnesses one from another. Popular feeling was so bitter against these men that they were often severely handled by the mob. Sometimes the mob made a mistake, and the innocent suffered for the guilty. In 1823 a coach containing an empty coffin was being drawn along the streets of Edinburgh; the people, suspecting that it was intended to convey a body, taken from some churchyard, seized the coach; it was with great difficulty that the police rescued the driver from the fury of the mob. The coach they could not save; it was taken through the streets, thrown over a mound, and smashed; the people then kindled a fire with the fragments, and danced round it. It turned out that the coffin was intended to convey to his house, in Edinburgh, the body of a physician who had died in the country.

On another occasion two American gentlemen, who were looking at the Abbey of Linlithgow after nightfall, were mistaken for resurrection-men, and assaulted by the mob.

One of the witnesses, called "A.B.," but who was probably Ben Crouch himself, stated that twenty-three in four nights was the greatest number he had ever obtained. He added, "When I go to work, I like to get those of poor people buried from the workhouses, because instead of working for one subject, you may get three or four. I do not think, during the time I have been in the habit of working for the schools, I got half a dozen of wealthier people." Another witness, who is called "C.D.," but who was, without doubt, the writer of the Diary, stated that, "according to my book," in 1809 and 1810 the number of bodies disposed of in England was 305 adults and 44 small; but the same year 37 were sent to Edinburgh, and the gang had 18 in hand, which were never used at all. In 1810-11, 312 adults were disposed of in the regular session, and 20 in the summer, in addition to 47 smalls. In the Report of the Committee in 1828, it was pointed out that, at that time, there were over 800 students attending the Schools of Anatomy in London, but of these not more than 500 actually worked at dissection. The number of subjects annually available for instruction amounted to between 450 and 500, or rather less than one for each student.

The average price of an adult body was
stated to be £4 4s. od. It may be here
explained that a "small" was a body under
three feet long; these were sold at so much
per inch and were generally classified as
"large small," "small," and "fœtus." The
earnings of the resurrection-men may be
gathered from the above entry. To take the
year 1810–11, the receipts for bodies alone
come to 1328 guineas; this is exclusive of
"smalls," and probably also of the teeth, in
which these men did a large trade. Teeth, in
those days, were very valuable; the amounts
received by some of the men for teeth only
will be dealt with in the chapter containing
biographical notices of some of the principal
London resurrection-men. It may be here
mentioned that on one occasion Murphy
obtained the entry to a vault belonging to
a meeting-house, on the pretence of selecting
a burial-place for his wife. Whilst in there
he managed to slip back some bolts, so that
he could easily gain an entrance at another
time; this he did at night, and got possession
of teeth by which he made £60.

From the statements of the teachers it is

most likely that £4 4s. od. is under the average price paid for bodies.   It must be remembered, too, that this amount does not include
the retaining-fee paid at the beginning of the
session, nor the "finishing-money" which was
demanded at its close.   The 1328 guineas
spoken of above would be divided amongst
six or seven persons, and this, for men in their
position, was a large income.   The biographical
notes of the chief workers in this horrible trade
will show that some few of them did save
money.   Taking them, however, as a whole,
they were a dissolute and ruffianly gang ; reference to the Diary proves their drunken habits,
and there is more than one entry to show that
they were often in pecuniary difficulties ; so
much so that on one occasion they were obliged
to have recourse to Mordecai, the Jew.

It was quite useless for those who had just
buried a relative or friend to depend either
upon the custodian of the burial-ground, or
upon the watch, to see that the newly-made
grave was not violated.   The resurrection-
men often met with a guard, instituted by the
friends of the deceased, who would take it in
turns to watch by the grave-side through the

whole night; these friends were frequently armed, and were not afraid to use their arms if the resurrection-men gave them an opportunity. As a rule the body-snatchers made off when they found a guard in the cemetery; it was to their interest not to create a riot, and if they were strong enough to drive off the watchers, the latter could soon raise a tumult, whereby the bodily safety of the thieves would be endangered.

Matters did not always pass off so peaceably, particularly in Ireland, as the following extract from an Irish newspaper for 1830 shows:

"DESPERATE ENGAGEMENT WITH BODY-SNATCHERS.—The remains of the late Edward Barrett, Esq., having been interred in Glasnevin churchyard on the 27th of last month (January), persons were appointed to remain in the churchyard all night, to protect the corpse from 'the sack 'em-up gentlemen,' and it seems the precaution was not unnecessary, for, on Saturday night last, some of the gentry made their appearance, but soon decamped on finding they were likely to be opposed. Nothing daunted, however, they returned on Tuesday morning with augmented force, and well armed.

About ten minutes after two o'clock three or four of them were observed standing on the wall of the churchyard, while several others were endeavouring to get on it also. The party in the churchyard warned them off, and were replied to by a discharge from fire-arms. This brought on a general engagement; the sack 'em-up gentlemen fired from behind the churchyard wall, by which they were defended, while their opponents on the watch fired from behind the tomb-stones. Upwards of 58 to 60 shots were fired. One of the assailants was shot—he was seen to fall; his body was carried off by his companions. Some of them are supposed to have been severely wounded, as a great quantity of blood was observed outside the churchyard wall, notwithstanding the ground was covered with snow. During the firing, which continued for upwards of a quarter of an hour, the church bell was rung by one of the watchmen, which, with the discharge from the fire-arms, collected several of the townspeople and the police to the spot— several of the former, notwithstanding the severity of the weather, in nearly a state of nakedness; but the assailants were by this

MORTSAFE IN GREYFRIARS CHURCHYARD, EDINBURGH.

time defeated, and effected their retreat. Several of the head-stones bear evident marks of the conflict, being struck with the balls, &c."

Most of the disgraceful riots which took place in the burial-grounds, were not between resurrection-men and friends guarding a grave, but between two gangs of body-snatchers. In cases of this kind one gang would do all in its power to bring its rival into disrepute ; the stronger party, after driving the weaker one away, would put the burial-ground into a most disgraceful state, and then give information against their opponents.

Besides watching, many other devices were tried to prevent the depredations of the resurrection-men ; spring guns were set in many of the cemeteries, but these were often rendered harmless. If the men intended going to a certain grave at night, late in the afternoon a woman, in deep mourning, would walk round the part of the cemetery in which the grave was situated, and contrive to detach the wires from the guns. Loose stones were placed on the walls of the grave-yard, so as to make scaling the walls almost an impossibility ; this

was useless when the custodian had a house
with a window looking into the burial-place.
If entrance could not be obtained in this way,
there was generally some other house through
which the men could gain admission to the
grave-yard. Mort-safes, or strong iron guards,
were placed over newly-made graves for protection; some of these can be seen at the
present day in the Greyfriars Churchyard,
Edinburgh (see illustrations).

Iron coffins were also used by some persons
to protect their friends from the Resurrectionist.
The following interesting advertisement appeared in *Wooler's British Gazette* for
October 13th, 1822:

"Many hundred dead bodies will be dragged
from their wooden coffins this winter, for
the anatomical lectures (which have just
commenced), the articulators, and for those
who deal in the dead for the supply of the
country practitioner and the Scotch schools.
The question of the right to inter in
iron is now decided. Lord Chief Justice
Abbott declared he wished they might be
generally used; Justice Bailey declared that
if the Ecclesiastical Court was to grant a suit

MORTSAFE IN GREYFRIARS CHURCHYARD, EDINBURGH.

for a fee, they, the Court of King's Bench, would grant a prohibition, knowing it had no such right. Sir William Scott, now Lord Stowell, decided and directed the interment without any extra fee, as this question was raised by an undertaker; those undertakers who have IRON COFFINS must divide the profits of the funeral with EDWARD LILLIE BRIDGMAN. TEN GUINEAS reward will be paid on the conviction of any Parish Officer demanding an extra fee, whereby I shall lose the sale of a coffin. The violation of the sanctity of the grave is said to be needful, for the instruction of the medical pupil, but let each one about to inter a mother, husband, child, or friend, say shall I devote this object of my affection to such a purpose; if not, the only safe coffin is Bridgman's PATENT WROUGHT-IRON ONE, charged the same price as a wooden one, and is a superior substitute for lead. Edward Lillie Bridgman, 34, Fish Street Hill, and Goswell Street Road, performs funerals in any part of the kingdom, and by attention to moderate charges insures the recommendation of those who employ him. Twenty-five private grounds within the Bills of Mortality receive them; dues

from seven shillings to one guinea. Patent
cast-iron tombs and tablets, superior to stone."

The advertisement is headed by a rough
cut, showing the coffin* and the iron clamps by
which it was fastened. There was another
maker of patent coffins, who is mentioned by
Southey in his ballad called *The Surgeon's
Warning*. The ballad represents the fear
of a dying surgeon, lest his apprentices should
serve him after death as he, during his life,
has served many other persons :

> "And my 'prentices will surely come
>   And carve me bone from bone,
>   And I, who have rifled the dead man's grave,
>   Shall never rest in my own.
>
> "Bury me in lead when I am dead,
>   My brethren, I entreat,
>   And see the coffin weigh'd I beg,
>   Lest the plumber should be a cheat.
>
> "And let it be solder'd closely down
>   Strong as strong can be, I implore,
>   And put it in a patent coffin
>   That I may rise no more.
>
> "If they carry me off in the patent coffin
>   Their labour will be in vain,
>   Let the undertaker see it bought of the maker,
>   Who lives in St. Martin's Lane."

* See illustration.

MANY Hundred DEAD BODIES will be dragged
from their Wooden Coffins this winter, to make
decent burials, (which have commenced on some
sites, and for those who died in the summer) by
the coroner practitioner, and the dead must be
reserved of right to later in iron, to keep them for the
Chief Justice Abbott declared the whether they were
generally bred. Justice Bailey declared that if the Ecclesiastical Court was so grievous and, and so they, the
Court of King's Bench would grant a prohibition, knowing it had no such right; and Mr. Justice per Lord
Powell, derided and divided the informant without any
extra fee, as this question was raised by an undertaker,
those outbreaks, who have IRON COFFINS made,
divide their quality to their funeral with EDWARD
LILLIE BRIDGMAN. TEN GUINEAS Reward will
be paid on the conviction of an parish officer demanding
an extra fee, whereby I shall lose the sale of a coffin.
The violation of the sanctity of the grave is said to be
needful, for the instruction of the medical pupil; but let
each one think to inter a mother, husband, child,
friend, say, shall I divide this object of my affections
such a purpose till cut; the only safe coffin is BRIDGMAN'S PATENT WROUGHT IRON one, charged the
same price as the cost, and is a superior substitute
for lead. Manufactory, 31, Fish Street Hill,
and Goswell Street Road... Funerals in any part
of the Kingdom, conducted on moderate charge...
invoice free, to those who employ him.
Twenty-five seven pounds within the bills of mortality,
remote there; from from 1s. to 21s. Patent cost may
now be paid to the quarter or more.

All the surgeon's wishes were duly carried out as regards his coffin ; money was also given to watchers to keep guard every night over the grave. The "'prentices," however, were able easily to buy the watchers, and so

> " They burst the patent coffin first,
>> And then cut through the lead,
>> And they laugh'd aloud when they saw the shroud,
>>> Because they had got at the dead.

> " And they allow'd the sexton the shroud
>> And they put the coffin back,
>> And nose and knees they then did squeeze,
>>> The surgeon in a sack.
>>> *　　　*　　　*　　　*
> "So they carried the sack pick-a-back,
>> And they carved him bone from bone,
>> But what became of the surgeon's soul,
>>> Was never to mortal known."

The following extract from a Scotch paper shows the alarm felt for the safety of the newly-buried :

" RESURRECTION-MEN.—Curiosity drew together a crowd of people on Monday, at Dundee, to witness the funeral of a child, which was consigned to the grave in a novel manner. The father, in terror of the resurrec-tion-men, had caused a small box, inclosing some deathful apparatus, communicating by means of wires, with the four corners, to be fastened on the top of the coffin. Immediately before it was lowered into the earth, a large quantity of gunpowder was poured into the box, and the hidden machinery put into a state of readiness for execution. The common opinion was, that if any one attempted to raise the body he would be blown up. The sexton seemed to dread an immediate explosion, for he started back in alarm after throwing in the first shovelful of earth."

Friends and relatives often placed objects on the newly-made grave, such as a flower or an oyster-shell, so that they might be able to tell if the earth had been disturbed. These objects were generally carefully noted by the resurrection-men, and were put back in their exact places after the body had been removed and the grave re-filled.

In some burial-grounds, houses were built in which the bodies could be kept until they were putrid, and therefore useless to the resurrection-men. Such a house is still standing in the burial-ground at Crail.*

As a rule, the resurrection-men were able

* See two following illustrations.

HOUSE AT CRAIL (Described on page 80). Ann. Dom. MDCCCXXVI."

Over the door is the following inscription: "Erected for securing the Dead.

HOUSE AT CRAIL (Described on page 89).

Over the door is the following inscription: "Erected for securing the Dead.   Ann. Dom. MDCCCXXVI."

not only to supply the London schools from the grave-yards in and around the Metropolis, but also to send bodies to some of the provincial schools ; the Diary shows that even Edinburgh received some of the proceeds of the work of this London gang. If, however, from increased vigilance or other causes, the supply of bodies ran short in London, recourse was had to the provinces. A case occurred some seventy years ago at Yarmouth. A man died, and was buried in St. Nicholas Church-yard. Not long after, his wife died also. On the husband's grave being opened, it was discovered that the man's body had been removed ; this led to a panic amongst people in Yarmouth who had recently buried friends in that churchyard. Many graves were opened, and, in a large number of instances, were found to have been violated. This led to a regular watch being established over newly-made graves in the churchyard. It was the custom of the resurrection-men, when they had bodies to send from the country to London, to forward them so that they should, in outward appearance, correspond with the class of goods exported from the place where the bodies had

been obtained. If the goods usually came to
London in crates, crates were used by the
body - snatchers ; if ordinary packing-cases,
then the bodies were enclosed in like recep-
tacles. The proceeds of the exhumations at
Yarmouth were probably packed in barrels,
and came through Billingsgate.

In 1826 three casks, labelled " Bitter Salts,"
were taken down to George's Dock at Liverpool, to be shipped on board the *Latona*,
bound for Leith ; a full description of this
transaction was printed as a broadside, of
which the following is a copy :

" RESURRECTIONISTS AT LIVERPOOL.

" Discovery of 33 Human Bodies, in Casks,
about to be shipped from Liverpool for Edinburgh, on Monday last, October 9, 1826.

" Yesterday afternoon, a carter took down
one of our quays three casks, to be shipped
on board the Carron Company's vessel, the
Latona, addressed to ' Mr. G. Ironson,
Edinburgh.' The casks remained on the quay
all night, and this morning, previous to their
being put on board, a horrible stench was
experienced by the mate of the *Latona* and

other persons, whose duty it was to ship them. This caused some suspicion that their contents did not agree with their superscription, which was ' Bitter Salts,' and which the shipping note described they contained. The mate communicated his suspicions to the agent of the Carron Company, and that gentleman very promptly communicated the circumstances to the police. Socket, a constable, was sent to the Quay, and he caused the casks to be opened, when Eleven Dead Bodies were found therein, salted and pickled. The casks were detained, and George Leech, the cart-man, readily went with the officer to the cellar whence he carted them, which was situated under the school of Dr. McGowan, at the back of his house in Hope Street; the cellar was padlocked, but, by the aid of a crow-bar, Boughey, a police officer, succeeded in forcing an entrance, and, on searching therein, he found 4 casks, all containing human bodies, salted as the others were, and three sacks, each containing a dead body. He also found a syringe, of that description used for injecting how tax into the veins and arteries of the dead bodies used for anatomization; he also

G

found a variety of smock-frocks, jackets, and trowsers, which, no doubt, were generally used by the Resurrectionists to disguise themselves. In this cellar were found twenty-two dead bodies, pickled and fresh, and in the casks on the quay, eleven, making in the whole thirty-three. The carter described the persons who employed him as of very respectable appearance, but he did not know the names of any of them.

" Information of the above circumstances was speedily communicated to his Worship, the Mayor, who sent for Dr. McGowan. This gentleman is a reverend divine, and teacher of languages ; he attended the Mayor immediately, and, in answer to the questions put to him, we understand he said, that he let his cellar in January last to a person named Henderson, who, he understood, carried on the oil trade, and that he knew nothing about any dead bodies being there. George Leech deposed that he plies for hire as a carter (the cart belongs to his brother) ; yesterday afternoon, between three and four o'clock, a tall, stout man asked him the charge of carting three casks from Hope Street to George's

Dock passage ; he replied, 2s. They then
went to Hope Street, where the witness found
two other men getting the first cask out of a
cellar under Dr. McGowan's schoolroom, and
witness assisted to get two other casks out of
the cellar ; the three were then put into his
cart, and the men who employed him gave him
a shipping note, describing the casks as containing ' Bitter Salts,' and told him to be
careful in laying them down upon the quay,
and that they were to be forwarded to Edinburgh by the *Latona*.

"Mr. Thomas Wm. Dawes, surgeon, of St.
Paul's Square, deposed that he had examined
the bodies, by the direction of the Coroner.
In one cask he had found the bodies of two
women and one man ; in another, two women
and two men ; in the third, three men and one
woman, and in the other casks and sacks he
found 22 (*sic*) bodies, viz., nine men, five boys,
and three girls; the bodies were all in a perfect
state ; those in the casks appeared to have
been dead six or seven days, and three men
found in the sacks appeared to have been dead
only three or four days. In each of the casks
was a large quantity of salt. There were no

external marks of violence, but there was a
thread tied round the toes of one of the
women, which is usual for some families to do
immediately after death. Witness had no
reason but to believe that they had died in a
natural way, and he had no doubt the bodies
had all been disinterred. The Season for
Lectures on Anatomy is about to commence
in the capital of Scotland.

" The police were ordered to be upon the
alert to discover the persons who had been
engaged in this transaction, but as yet nothing
further has been ascertained. The bodies, by
the direction of the Coroner, were buried this
morning in the parish cemetery, in casks, as
they were found.

" It is not yet ascertained whence these
bodies have been brought, but it is supposed
that the Liverpool Workhouse Cemetery has
been the principal sufferer. Some of them are
so putrid, that it is extremely dangerous to
handle them.   BOAG, PRINTER."

The statements in this broadside are quite
true, and agree with the account which is
to be found in the *Liverpool Mercury* for

October 13th, 1826. Henderson, who was a Greenock man, and the principal in this business, escaped, and could not be brought to justice ; but a man named James Donaldson, who was a party to the transaction, was made to pay a fine of £50, and was sent to Kirkdale Gaol for twelve months.

From Ireland very many bodies were exported, chiefly to Edinburgh ; a better price could be obtained there than in Dublin, and the consequence was that the Irish schools were often very badly supplied with subjects. In Dublin there were several ancient burial-grounds, all badly protected; the poor were all buried in one part, and, as their friends were generally unable to afford watchers, their bodies fell an easy prey to the resurrection-men. In January, 1828, the detection of a body about to be exported caused a tumult in the streets of Dublin, and led to the murder of a man named Luke Redmond, a porter at the College of Surgeons.* The body-snatchers in Dublin seem to have done more damage than the men engaged in a like

* CAMERON, *History of Roy. Coll. Surgeons in Ireland*, p. 113.

occupation in London ; they were not content with taking the bodies, but, in addition, they broke the tomb-stones, and played general havoc in the grave-yards.

According to the following cutting from the *Universal Spectator and Weekly Journal*, May 20th, 1732 (printed in *Notes and Queries*, 5th ser. i. 65), bodies were sometimes taken for other than dissection purposes. " John Loftas, the Grave Digger, committed to prison for robbing of dead corpse, has confess'd to the plunder of above fifty, not only of their coffins and burial cloaths, but of their fat, where bodies afforded any, which he retail'd at a high price to certain people, who, it is believed, will be call'd upon on account thereof. Since this discovery several persons have had their friends dug up, who were found quite naked, and some mangled in so horrible a manner as could scarcely be suppos'd to be done by a human creature."

Southey also refers to this in the poem before quoted, where he makes the surgeon say in his lamentation,

> " I have made candles of infants' fat."

CHAPTER III.

IT is well-nigh impossible to read of all these misdoings and not to ask why the Government did not step in and put a stop to them? It was urged by many that a short Act should be passed, making the violation of a grave a penal offence, as it was in France. There was a general agreement that anatomical education was absolutely necessary for medical men, and that this education was an impossibility without a supply of subjects; yet there was a great reluctance to interfere by legislation. The Home Secretary told a deputation that there was no difficulty in drawing up an effective Bill; the great obstacle was the prejudice of the people against any Bill; this impediment, he added, had not been trifling.

By no class of men was legislation more earnestly asked for than by the teachers of anatomy; to them the system then in vogue

was not only degrading, but it meant absolute
ruin.

There was at that time no property in a dead
body, and a prosecution for felony could not
take place unless some portion of the grave-
clothes or coffin could be proved to have been
stolen with the body.   The resurrection-men
were well aware of this fact, and generally took
precaution to keep themselves out of the
meshes of the law.

There had been some successful prosecutions
like that of Holmes and Williams before mentioned, but magistrates would not always
convict.

In 1788 this question first came before the
Court of King's Bench in the case of Rex *v.*
Lynn.   The indictment charged the prisoner
with entering a certain burial-ground, and
taking a coffin out of the earth, and removing
a body, which he had taken from the coffin, and
carrying it away, for the purpose of dissecting
it.   For the defence the following passage from
Lord Coke was quoted : " It is to be observed
that in every sepulchre that hath a monument
two things are to be considered, viz., the monument, and the sepulture or burial of the dead :

the burial of the cadaver is *nullius in bonis*, and belongs to Ecclesiastical cognizance ; but as to the monument, action is given at the common law for defacing thereof." The only Act of Parliament which was said to bear on the subject was that of 1 Jac. I., c. 12, which made it felony to steal bodies for purposes of witchcraft. The Court, however, held in this case of Rex *v.* Lynn that to take a body from a burial-ground was an offence at common law, and *contra bonos mores*. In the judgment it was stated that as the defendant might have committed the crime through ignorance, no person having been before punished for this offence, the Court only fined him five marks. The reference here, to no one having been previously punished for a like offence, refers only to the Superior Courts, as there had been convictions at the Police Courts and the Old Bailey. Despite this decision of the Court, prosecutions were very seldom undertaken, although Southwood Smith* states that there had been fourteen convictions in England during the year 1823. In examination before the Committee on Anatomy, in

* *Use of the Dead to the Living.*

1828, Mr. Twyford, one of the magistrates at Worship Street Police Court, stated that he had not had more than six cases in as many years.

The following account of proceedings at Hatton Garden Police Court, in 1814, will show the difficulty of getting a conviction. In this case there seems to have been no one to identify the bodies. It is very improbable that in a case of this sort the authorities of burial-grounds would come forward to give evidence, and so confess their own negligence.

"HATTON GARDEN.

"T. Light, W. Arnot, and —— Spelling, were brought up on Wednesday. It appeared that the prisoners were going up Holborn about half-past four o'clock on Tuesday afternoon, with a horse and cart; they were observed by two officers, who, knowing the prisoners to be resurrection-men, stopped the horse and cart, and, after a hard contest, succeeded in securing the prisoners. They then examined the contents of the cart, and found it contained seven dead bodies of men and women; one of the bodies was headless, but how it came to be so remains as yet to be cleared up. They were packed up in bags and baskets. The prisoners were followed by an immense crowd to Hatton Garden Office, whence they were committed to prison, and the bodies deposited in the lock-up house. The cart was hired at Battle Bridge. Some of the officers were sent to make enquiry at the different burying-grounds. The Office was crowded with men and women, who had some of their relatives buried on Sunday last, to see if they could recognize any of the bodies. They were brought up again on Thursday, and discharged."

In 1822 the case of Rex *v.* Cundick was tried at Kingston Assizes, *coram* Graham.* This was an indictment for misdemeanour. A man named Edward Lee was executed in the parish of St. Mary, Newington ; George Cundick was employed by the keeper of the gaol to bury the body of Lee, and for this he was paid. Instead of burying the corpse, he sold it for dissection, or, in the words of the indictment, he "for the sake of wicked lucre and gain did take and carry away the said body,

* *D. and R. Nisi Prius Repts.* i. 13.

and did sell and dispose of the same for the purpose of being dissected, cut in pieces, mangled, and destroyed, to the great scandal and disgrace of religion, decency, and morality, in contempt of our Lord the King, and his laws, to the evil example of all other persons in like cases offending." The evidence showed plainly that Cundick had had possession of the body, and that he had received the burial fees. On the friends of Lee wishing to see the corpse, Cundick declared that it was already buried ; but several days after this he clandestinely went through the ceremony of burying a coffin filled with rubbish. It was also proved that Cundick had been seen to remove a heavy package from his house at night, and that the body of Lee had been identified in a dissecting-room. The defence was, in the first place, that the indictment was bad "as a perfect anomaly in the history of criminal pleading." In the econd place, if the indictment were good, it was unsupported by evidence. It was argued by counsel that the only evidence before the Court was that the body was not buried, and that it was found at a dissecting-room. Without the production of the owner of the dissecting-room, and the proof that he had bought the body from Cundick, the jury could not be asked to give a verdict against the defendant. The Judge, however, over-ruled these objections, and the jury found the prisoner guilty.

These trials and verdicts made it still more difficult than before to get subjects for dissection, as even men of the Resurrectionist class hesitated to run the risk of getting the punishment, which now the superior Courts had upheld. Those who did run this risk very naturally expected a price proportionate to the danger, and so the cost of subjects was still more increased.

But to surgeons, and to teachers of anatomy, by far the most important trial of all was that of John Davies and others, of Warrington, for obtaining the body of Jane Fairclough, which had been taken from the chapel-yard belonging to the Baptists, at High Cliff, Appleton, Cheshire, in October, 1827. This case was tried at Lancaster Assizes, March 14th, 1828. The defendants were John Davies (a medical student at the Warrington Dispensary), Edward Hall (a surgeon and apothecary in practice at Warrington), William Blundell

(an apprentice to a stationer in the same town),
and Richard Box. Thomas Ashton was also
included in the indictment, but no evidence
was offered against him. There were fourteen
counts in the indictment, ten charging the
defendants with conspiracy, and four charging
them with unlawfully procuring and receiving
the body of Jane Fairclough. It appears, from
the report of the trial, that Davies called on
Dr. Moss, one of the Physicians to the Dispensary, and obtained permission to use a
building in his garden for the purpose of
dissecting a subject which he had purchased.
Mr. Hall, on behalf of Davies, paid four
guineas to the men who brought the body
to a cellar in Warrington, but he knew nothing
more of the transaction ; from the cellar the
body was removed to Dr. Moss' premises by
Blundell and another man, and was received by
Davies and a servant of Dr. Moss. Information of the exhumation seems to have quickly
got about. The funeral was on a Friday ; on
the Monday following the grave was undisturbed, but on Tuesday the soil was spread
about, and an examination of the grave showed
that the corpse had been removed. The body

was identified at Dr. Moss' house, and was taken away before any dissection had been performed on it.

In charging the jury, Mr. Baron Hullock said that, as conspiracy was an offence of serious magnitude, they should be satisfied, before finding a verdict of guilty on the former part of the indictment, that the conduct of the defendants was the result of previous concert. . . . . If any of the defendants were in possession of the body under circumstances which must have apprized them that it was improperly disinterred, the jury would find them guilty of the latter part of the charge. The only bodies legally liable to dissection in this country were those of persons executed for murder. However necessary it might be, for the purposes of humanity and science, that these things should be done, yet, as long as the law remained as it was at present, the disinterment of bodies for dissection was an offence liable to punishment. The jury found all the defendants not guilty of the charge of conspiracy, but they pronounced Davies and Blundell guilty of possession of the body, with knowledge of the illegal disinterment. The defendants were brought up

for judgment in London in May, 1828.   Mr.
Justice Bayley, in passing sentence, said that
"there were degrees of guilt, and in this case
the defendants were not the most criminal
parties."   He sentenced Davies to a fine of
£20, and Blundell to a fine of £5.

It will be noted that in this trial there is no
charge against anyone for violating the grave,
or stealing the body.   The fines were inflicted
on Davies and Blundell for having the body in
their possession, knowing it to have been
disinterred.   This decision, therefore, as before
stated, was of the utmost importance to teachers
of anatomy, as they were clearly liable to
punishment for all the subjects supplied to
them by the Resurrectionists.   The teachers
knew well the sources from which the bodies
were obtained, and were only driven to get
them in the way they did through there being
no regular supply of subjects from a legitimate
source.   The feeling that legislation on this
subject was absolutely necessary, was more
keenly felt than ever, and the teachers did all
they could to get a change in the laws.   Many
pamphlets were issued from the press, urging
this duty upon Parliament ; it was pointed out

Surgical Operations, or a New method of O[c]ta-x-ent Subjects.

that if a supply of bodies could be regularly
obtained in a legal way, the trade of the Resurrectionist would at once cease. There were
many who doubted this, but subsequent
events proved the statement to be strictly
accurate.

It was very strongly urged that the Act of
Geo. II., which ordered the bodies of all
murderers executed in London and Middlesex
to be anatomized by the Surgeons' Company,
ought to be repealed. No doubt this provision
much increased the dislike of the poor to any
regulations by which the bodies of their friends
might be given up for dissection after death.
It was felt that dissection by the Surgeons was
part of the sentence passed on a murderer,
and therefore carried with it shame and
disgrace. To make provision by law, therefore,
for the dissection of the bodies of any other
class of persons was, not unnaturally, distasteful,
in that it partly put them in the same position
as murderers.

The answer to the desire for the repeal of
this obnoxious clause was that nothing must
be done to weaken the law ; it was stated that
to withdraw the part of the sentence which

related to dissection would rob the punishment
of its prohibitive effect. It is somewhat difficult to understand the argument; surely if the
risk of suffering the extreme penalty of the law
would not keep a man from crime, the extra
chance of being dissected after death could
hardly be expected to do so. As Sir Henry
Halford said, "I certainly think that while that
law remains they [the public] will connect the
crime of murder with the practice of dissection;
an order to be dissected, and a permission to be
dissected, seem to be too slight a distinction."

Another objection to the dissection of murderers came from the teachers. They stated
that when the body of a notorious criminal was
lying at either of the Anatomical Schools, the
proprietor was pestered by persons of a morbid
turn of mind for permission to view the body.
This difficulty was also felt by the College of
Surgeons, and in consequence a placard was
hung up outside the place where the dissections
were made, giving notice that no person could
be admitted, unless accompanied by a member
of the Court of Assistants.

To make dissection less distasteful to the
general public, and to show the advantages of

anatomy, some endeavours were made to explain the structure of the human body to non-professional persons. In Ireland Sir Philip Crampton lectured with open doors, and gave demonstrations in anatomy to poor people. These persons, he tells us, became interested in the subject, and often brought him bodies for dissection. A newspaper cutting of 1829 shows that this was also tried in London. A surgeon called in the overseers and church-wardens of St. Clement Danes, and gave a demonstration on a body, explaining its con-struction, and the use of the internal organs. "By this means," says the paragraph, "he so fully absorbed the self-interest of his audience as to extinguish the pre-conceived notions of horror and disgust attached to the idea of a spectacle of this description. The enlightened governors of the parish assented to the *post mortem* examination of the body of every unclaimed pauper, an enquiry into whose case might appear conducive to the interests of medical science."

It has been already pointed out that, to try to overcome the repugnance to dis-section, some persons left specific instructions

that their bodies should be used for this
purpose.

The representations of the teachers were so
far successful, that in 1828 a Select Committee
was appointed by the House of Commons "to
enquire into the manner of obtaining subjects
for dissection in the Schools of Anatomy, and
into the state of the law affecting the persons
employed in obtaining and dissecting bodies."
Amongst those who gave evidence before the
Committee were the principal teachers of
anatomy, and three of the resurrection-men.
The tone of the Report was decidedly in
sympathy with the teachers, but it strongly
condemned the way in which they were compelled to obtain bodies for dissection. After
showing how badly off English students were
for opportunities of learning anatomy, as compared with those of foreign countries, and
pointing out that those students who really
wished to master their art were compelled to
go abroad, the Report proceeds : " These disadvantages affecting the teachers are such,
that except in the most frequented schools,
attached to the greater hospitals, few have
been able to continue teaching with profit, and

some private teachers have been compelled to give up their schools. To the evils enumerated it may be added, that it is distressing to men of good education and character to be compelled to resort, for their means of teaching, to a constant infraction of the laws of their country, and to be made dependent, for their professional existence, on the mercenary caprices of the most abandoned class in the community."

In March, 1829, Mr. Warburton obtained leave to introduce into the House of Commons "A Bill for preventing the unlawful disinterment of human bodies, and for regulating Schools of Anatomy." In this Bill it was enacted that persons found guilty of disinterring any human body from any churchyard, burial-ground or vault, or assisting at any such disinterment, should be imprisoned for a term not exceeding six months for the first offence, and two years for the second offence. Seven Commissioners were to be appointed ; the majority of these were not to be either physicians, surgeons, or apothecaries. All unclaimed bodies of persons dying in work-houses or hospitals, were, seventy-two hours

after death, to be given over for purposes of
dissection; but if within this specified time a
relative appeared and requested that the body
might not be used for anatomical purposes,
such request was to be granted.   Another
proposed change in the law was that a person
might legally bequeath his body for dissection;
in such cases the executors, administrators, or
next-of-kin had the option of carrying out the
wishes of the testator, or declining to do so, as
they thought fit.   A heavy penalty was laid on
persons who were found carrying on human
anatomy in an unlicensed building, and it was
made an offence to move a body from one
place to another, without a licence for so
doing.   All bodies used for dissection were to
be buried; the penalty for failing to do this
was fifty pounds.

One great blot on this Bill was the neglect-
ing to repeal the clause which ordered the
bodies of murderers to be given up for dissection.   As pointed out on page 87, this was
one of the great reasons which made dissection
so hateful to the poor.   During the debate, a
motion was made by Sir R. Inglis "to repeal
so much of the Act 9 Geo. IV. cap. 31, as

empowers judges to order the bodies of murderers to be given over for dissection." This, however, was lost, eight members only voting for the amendment, and forty against.

There was strong opposition to the Bill outside the House. Some of the private teachers were very uneasy as regarded the effect of the Bill on themselves. The measure spoke of "recognized teachers" and "hospital schools," and all those who were to be entitled to the benefits of the Act were to have licences from one of the Medical Corporations. The proprietors of the smaller schools felt that this would result in their extinction, and that the teaching would all pass to the large schools. In the country, too, there was strong opposition to the Bill, as practitioners there felt that they were excluded from any benefit. The *Lancet*, always ready in those days with a nickname, dubbed the measure "A Bill for Preventing Country Surgeons from Studying Anatomy." The College of Surgeons also petitioned against the Bill. The Council felt that the appointment of Commissioners, who were to have complete control over all schools and places of dissection, would greatly interfere

with the privileges of the College. It was
pointed out to the House of Commons that
the establishment of a Board, such as that proposed by the Bill, was virtually placing the
whole profession of surgery under the control
of Commissioners, not one of whom need be a
member of the profession, and the majority of
whom must not be so.

Another fault of the Bill was that it did not
apply to Ireland. A large supply of bodies
was regularly sent from that country to
England and Scotland, and it was felt that to
exclude Ireland from the provisions of the Bill,
was simply increasing the temptation for bodies
to be still more largely exported therefrom.

It was also argued that the Bill would tell
hardly against the poor, as they would refuse
to go into workhouses or hospitals if they
thought that their bodies would be dissected
after death. For this objection there was no
foundation, and Mr. Peel pointed out, in the
debate on the third reading, that "it was the
poor who would really be benefited by the
measure. The rich could always command good
advice, whilst the poor had a strong interest
in the general extension of anatomical science."

The Bill passed the Commons, but was lost in the Lords.

In 1830, Lord Calthorpe was to have again introduced the Bill into the Upper House, but the intention was abandoned on account of the threatened dissolution of Parliament. As the *Lancet* expressed it, " Dissolution has so many horrors, that a discussion on the *subject* at the present time would be by no means agreeable."

Public feeling was now very strong in favour of some law to prevent the wholesale spoliation of graves, which was going on practically unchecked. But, as has happened frequently in legislation, the absolute necessity for a change in the law was brought within the range of practical politics by a crime of a most diabolical character, one which, in this country, created a sensation equal to that raised in Scotland by the atrocities of Burke and Hare in Edinburgh.

On November 5th, 1831, two men, named Bishop and May, called at the dissecting-room at King's College, and asked Hill, the porter, if he "wanted anything." On being interrogated as to what they had to dispose of, May replied, " A boy of fourteen." For this body they asked 12 guineas, but ultimately agreed to bring it in for 9 guineas. They went off, and returned in the afternoon with another man named Williams, *alias* Head, and a porter named Shields, the latter of whom carried the body in a hamper. The appearance of the subject excited Hill's suspicion of foul play, and he at once communicated with Mr. Partridge, the Demonstrator of Anatomy. A further examination of the body by Mr. Partridge confirmed the porter's suspicions.* To delay the men, so that the police might be communicated with, Mr. Partridge produced a £50 note, and said that he could not pay until he had changed it. Soon after, the police officers appeared upon the scene, and the men were given into custody. At the coroner's inquest a verdict of " Wilful murder against some person or persons unknown" was brought in, the jury adding that there was strong suspicion against Bishop and Williams. The prisoners were not allowed to go free, but were kept in custody. Bishop, Williams, and May were tried at the Old Bailey, December, 1831. The evidence given against them showed that

* See also page 56.

they had tried to sell the body at Guy's Hospital ; being refused there, they tried Mr. Grainger, at his Anatomical Theatre, but with no success. Then they tried King's, where their crime was detected. The body was proved to be that of an Italian boy, named Carlo Ferrari, who obtained his living by showing white mice. The boy's teeth had been extracted, and it was proved that they had been sold by one of the prisoners to Mr. Mills, a dentist, for twelve shillings. The jury found all three prisoners guilty, and they were sentenced to death.

From the subsequent confessions of Bishop and Williams, it was shown that they had enticed the boy to their dwelling in Nova Scotia Gardens ; there they drugged him with opium, and then let his body into a well, where they kept it until he was suffocated. To the last the prisoners declared that the deceased was not the Italian boy, but a lad from Lincolnshire. They seem to have had great difficulty in disposing of the body, as Bishop, in his confession, said that, before taking it to Guy's, they had tried Mr. Tuson and Mr. Carpue, both in vain. Bishop and Williams confessed, also, to the murder of a woman named Fanny

Pigburn, and a boy, whose name was supposed
to be Cunningham.   Both of these bodies they
sold for dissection.   May was respited, and
was sentenced to transportation for life.   On
hearing of his respite, May went into a fit, and
for some time his life was despaired of; he,
however, partially recovered, but his feeble
state of health was aggravated by the annoy-
ance he received from the other convicts on
board the hulks.   He died on board the
Grampus in 1832.

May can hardly be described as even a
minor poet, if the following verse, written
whilst in prison, may be taken as a fair
sample of his compositions :

> "James May is doomed to die,
>   And is condemned most innocently ;
>   The God above, He knows the same,
>   And will send a mitigation for his pain."

At the execution of Bishop and Williams,
there was a scene of the most tremendous
excitement.   By some mistake, three chains
hung from the gallows ; one was taken away
as soon as the error was noticed, and this was
recognized by the crowd as a sign that May
had been reprieved.

The *Weekly Dispatch* sold upwards of 50,000 copies of the number which contained the confessions of the murderers. Many persons were injured in the crowd, and the *Dispatch* states that those who were hurt were attended to " by Mr. Birkett, the dresser to Mr. Vincent, who had been in attendance [at St. Bartholomew's Hospital] to receive any accident that might be brought in."

Bishop was the son of a carrier between London and Highgate, and on the death of his father he succeeded to the business. This he soon sold, and became an informer. He got mixed up with some of the resurrection-men, and then regularly took to the occupation. Williams, *alias* Head, was Bishop's brother-in-law, and was a well-known member of the resurrection-gang.

In the *Weekly Dispatch* for December 11th, 1831, the following curious information respecting Williams appeared :

"EXCISE COURT.—YESTERDAY.

"THE KING *v.* THOMAS HEAD, *alias* WILLIAMS, THE MURDERER.—The Court was occupied during a great part of the morning in hearing the evidence in the case of Head, *alias* Williams (who was hung with Bishop) for carrying on an illicit trade in the manufacture of glass. It appeared that the deceased was a *Cribb Man*, or regular porter, to private glass blowers. There were found on the premises at No. 2, Nova Scotia Gardens (the scene of the late murders), a regular furnace, and all the necessary apparatus for the manufacture of glass, which trade it appears was carried on to a very considerable extent on the premises. Alexander M'Knight, an officer of Excise, deposed that on the 6th of August last, he went to No. 2, Nova Scotia Gardens, and made a seizure of 68 cwt. of manufactured glass, 24 cwt. of cullet, and 16 cwt. of iron, articles used in the manufacture of glass. In about half-an-hour afterwards he saw Williams come out of Bishop's yard ; Williams spoke to witness, and called him by an opprobrious name for having made the seizure. Judgment 'abated,' the goods to be returned to the Excise Office to be condemned."

May had been brought up as a butcher, but this trade he gave up, and became possessed of a horse and cart with which he was supposed to

JOHN HEAD, *alias* THOMAS WILLIAMS. JOHN BISHOP.
Executed December 5, 1831. From Drawings by W. H. Clift, made directly after the execution.

ply for hire. The real business of the vehicle, however, seems to have been to convey bodies from place to place for the Resurrectionists. Shields, the porter to the gang, had been watchman and grave-digger at the Roman Catholic Chapel in Moorfields, so that he was most useful to the other Resurrectionists in giving information, and in granting facilities for the removal of bodies. No evidence was offered against him in connection with the murder of the Italian boy. Soon after the trial he attempted to get work as a porter in Covent Garden Market, but on his being recognized by those working there, a shout of " Burker!" was raised, and Shields narrowly escaped with his life, and took refuge in the Police Office.

This one incident as regards Shields gives an idea of the public feeling towards the resurrection-men, and that feeling was quite as bitter towards the anatomists. It was therefore absolutely necessary that some determined steps should be taken as regards legislation.

In December, 1831, Mr. Warburton again introduced a Bill into the House of Commons; it passed safely through both Houses, and

became law on August 1st, 1832. By this new
Act the Secretary of State for the Home
Department in Great Britain, and the Chief
Secretary in Ireland, were empowered to grant
licences for anatomical purposes to any person
lawfully qualified to practise medicine, to any
professor or teacher of anatomy, and to
students attending any school of medicine, on
an application signed by two justices of the
peace, who could certify that the applicant
intended to carry on the practice of anatomy.
It was enacted that executors, or other persons
having lawful possession of a body (provided
they were not undertakers, or persons to whom
the body had been handed over for purposes of
interment), might give it up for dissection
unless the deceased had expressed a wish
during his life that his body should not be so
used, or unless a known relative objected to
the body being given up. If a person had
expressed a wish to be dissected, this wish was
to be carried out unless the relatives raised any
objection. No body might be moved for
anatomical purposes until forty-eight hours
after death, nor until the expiration of a twenty-
four hours' notice to the Inspector of Anatomy ;

a proper death certificate had also to be signed by the medical attendant before the body could be moved. Provision was made for the decent removal of all bodies, and for their burial in consecrated ground, or in some public burial-ground in use for persons of that religious persuasion to which the person, whose body was so removed, belonged. A certificate of the interment was to be sent to the Inspector within six weeks after the day on which the body was received. No licensed person was to be liable to any prosecution, penalty, forfeiture, or punishment for having a body in his possession for anatomical purposes according to the provisions of the Act.

Perhaps the most important clause was that which did away with the dissection of the bodies of murderers. This was done by Section XVI., which ran as follows :

"And whereas an Act was passed in the Ninth Year of the Reign of His late Majesty, for consolidating and amending the Statutes in England relative to Offences against the Person, by which latter Act it is enacted, that the Body of every Person convicted of Murder shall, after Execution, either be dissected or hung in Chains, as to the Court which tried the Offender shall deem meet; and that the Sentence to be pronounced by the Court shall express that the Body of the Offender shall be dissected or hung in Chains, whichever of the Two the Court shall order. Be it enacted, That so much of the said last-recited Act as authorizes the Court, if it shall see fit, to direct that the Body of a Person convicted of Murder shall after Execution, be dissected, be and the same is hereby repealed: and that in every case of Conviction of any Prisoner for Murder, the Court before which such Prisoner shall have been tried shall direct such Prisoner either to be hung in Chains or buried within the Precincts of the Prison in which such Prisoner shall have been confined after conviction, as to such Court shall deem meet; and that the sentence to be pronounced by the Court shall express that the body of such Prisoner shall be hung in Chains, or buried within the Precincts of the Prison, whichever of the two the Court shall order."

Three Inspectors were appointed to carry out the provisions of the Act. The first Inspectors were Dr. J. C. Somerville, for

England; Dr. Craigie, of Edinburgh, for Scotland; and Sir James Murray, of Dublin, for Ireland. There was no provision for punishing persons found violating graves; it had been already decided that this was an offence at common law; and presumably the framers of the Act had, at last, sufficient faith in their measure to believe that it would put an end to the proceedings of the resurrection-men. If that were so, they were not disappointed. After the passing of the Act the resurrection-man, as such, drops out of history; his occupation was gone, and one of the most nefarious trades that the world has ever seen came completely to an end. Public feeling against these men did not all at once subside; this strongly militated against their getting employment, and some of them moved to other quarters, where they lived under assumed names.

In looking back it is impossible not to regret that Parliament was so slow to believe that legislation in the direction of the Anatomy Act would do away with the evils of the resurrection-men. This fact was urged upon them by the teachers; but popular feeling was so

dead against the anatomists, who were thought
to be responsible for even the worst crimes of
the resurrection-men, that Parliament seemed
to fear to do anything which favoured the
teachers, although the great disadvantages
under which they suffered were thoroughly
well known.   Perhaps the best tribute to the
success of the Act, is the very small alter-
ations which have been made in it between
1832 and the present day.

A glance at the regulations in force in
foreign countries for the supply of bodies,
at the time of the passing of the Anatomy
Act, shows that when a fair provision was
made by law for the supply of bodies,
the resurrection-men were unknown.   The
great advantages of the student on the
Continent, as compared with his brethren in
England, were thus pointed out to the Committee by Mr. [afterwards Sir] William Law-
rence: "I see many medical persons from
France, Germany, and Italy, and have found,
from my intercourse with them, that anatomy
is much more successfully cultivated in those
countries than in England ; at the same time I
know, from their numerous valuable publica-

tions on anatomy, that they are far before us in this science; we have no original standard works at all worthy of the present state of knowledge." It was also shown that this fact was chiefly the result of the greater opportunities for getting subjects abroad, and that teachers found that those English students who had been to foreign schools were the best informed.

Before the Revolution in France the hospitals of Paris were supported by voluntary contributions, and each had separate funds and Boards of Management, similar to the hospitals in London at the present day. At the Revolution these Boards were consolidated, and one administrative body was formed. This "Administration des Hôpitaux, Hospices et Secours à Domicile de Paris," carried into effect the law passed by the Legislative Assembly, that the bodies of all those persons who died in hospitals, which were unclaimed within twenty-four hours after death, should be given up for anatomical purposes. The distribution from the hospitals to the medical schools was systematically carried out, generally at night.

By Art. 360 of the Penal Code, the punishment
for violation of a place of sepulture was
imprisonment for a term varying from three
months to a year, and a fine of from 60 to 200
francs. The result of these regulations was
that exhumation for anatomical purposes was
quite unknown.

In Germany the bodies of persons who died
in prisons, or penitentiaries, and those of suicides,
were given up for dissection, unless the friends
of the deceased cared to pay a certain sum to
the funds of the school ; in this case the body
was handed over to the friends. Other sources
of supply were the bodies of those persons who
died without leaving sufficient to pay the cost
of burial, poor people who had been supported
at the public cost, all persons executed, and
public women. Although these regulations
were not rigorously carried out, there was
an ample supply of bodies for anatomical
purposes, and the resurrection-men were unknown.

In Austria, if the medical attendant thought
necessary, a *post mortem* was made on all

patients who died in hospital, but only unclaimed bodies were used for dissection; these
were given up to the teachers forty-eight hours
after death. In Vienna the supply came from
the General Hospital; this was sufficient for
all purposes, and there was no recourse to
exhumation.

The supply in Italy came from a source
similar to that of the other countries named.
The rule was that all bodies of persons who
died in hospital were given up for dissection if
required; but, by paying the cost of the funeral,
friends could, if they wished, take away the
body. This, however, was seldom done.
There was generally a sufficient supply of
bodies; but, if this ran short, the subjects
were obtained from "the deposit" of poor
people who died and were buried at the
public cost. In every parish church in Italy
there was a chamber in which all the dead
bodies of the poor were deposited during the
day-time, after the religious ceremonies had
been performed over them in the church; at
night these bodies were removed either to the
dissecting-room or to the burial-fields, outside

the town.   Body-snatching was quite unknown.

There was an ample supply of bodies in
Portugal from similar sources.   Mortality was
very high amongst infants, who were put into
roda, or foundling cradles ; the bodies of these
children could be obtained without any difficulty.   In Portugal the resurrection-man did
not exist.

In Holland there was no lack of material for
teaching anatomy, and for students to learn
operative surgery on the dead body.   The
Dissecting School at Leyden was supplied
from the civil hospitals at Amsterdam.
There was no prejudice against dissection
in Holland ; in all the principal towns lectures
on anatomy were publicly given, and dissected
subjects were exhibited.   Here, again, exhumation was not necessary, and was unknown.

In the United States the laws relating to
anatomy varied very considerably in the
different States ; there was no regular supply
for the schools, and, consequently, subjects had

to be obtained by the aid of resurrection-men. In Philadelphia and Baltimore, the two great Medical Schools of the United States in those days, the supply of bodies was obtained almost entirely from the " Potter's Field," the burial-place of the poorest classes. This exhumation was carried on by an understanding with the authorities that the men employed by the schools in this work should not be interfered with. Dissection in the United States was, as in this country, looked upon with great aversion ; this was, no doubt, mainly owing to the fact that the bodies used for this purpose were obtained from the graves.

CHAPTER IV.

THE Diary of a Resurrectionist is written on 16 leaves, but is, unfortunately, imperfect. The first entry is November 28th, 1811, and the last December 5th, 1812. There are no entries in May, June, and July; during these months there would be little demand for subjects, as the sessions of the Anatomical Schools ran from October to May. Besides this, the light nights would interfere with the work of the men. The entry under the date February 25th refers to this: "the moon at the full, could not go." The state of the moon was of great importance to these men in their work; the writer of the Diary has on one of the pages copied out the "Rules for finding the moon on any given day," and has set out the epact for 1812 and 1813.

There is no clue in the Diary itself as to the name of the writer, and, unfortunately, Sir Thomas Longmore* was quite unable to

* See page vi.

remember the name of the individual from whom he received it. Feeling was very strong against the men who had been engaged in the resurrection business, and therefore, when information was required from them, every effort was made to keep their names secret. As late as 1843, when the *Life of Sir Astley Cooper* was published, the name of this man was carefully concealed, though most of the other members of the gang were freely spoken of under their full names. Bransby Cooper* quotes a written statement made by this man to the effect that he was in Maidstone Gaol in October, 1813. Enquiry at the gaol has, however, failed to find any mention of him; the original document is not forthcoming, and it is very probable that there is a mistake as regards the date. In this statement he is called Josh. N——, and Bransby Cooper speaks of him as N. There is a letter on "Body-snatchers" in the *Medical Times*, 1883, vol. i. p. 343, signed, "Your Old Correspondent"; the writer of the letter was, in all probability, Mr. T. Madden Stone, who had been a correspondent of the journal in

* *Life of Sir Astley Cooper*, vol. i. p. 422.

question from the time of its foundation. Mr. Stone had a valuable collection of papers and autographs, and his letter is really a reprint of a paper in his possession relating to payments made to the resurrection-men. In it occurs the following passage: "N.B., Sir Astley Cooper great friend to Naples." Mr. Stone presented a large number of papers and letters to the Royal College of Surgeons, but this particular one is not in the collection. It is curious that Bransby Cooper makes no special mention of Naples in his book, although he gives an account of all the other men with whom Sir Astley had any dealings. He gives a long notice of " N.," and mentions that he wrote the Diary from which quotations are made; this is the document now under consideration.

The witness " C. D.," who was examined before the Committee on Anatomy in 1828, was, in all probability, Naples; he gave statistics to show the number of bodies obtained, and stated that the figures were taken "from my book." The letters "C. D." are not given as initials; the three resurrection-men who gave evidence were distinguished as "A.B.," "C.D.," and "F.G." The testimony was probably given on the condition that no names were revealed, and, therefore, definite information cannot be obtained as to " C. D.'s " real name from the House of Commons.

On one page of the Diary is written " Miss Naples." This does not prove much, as the names of several other females are mentioned; not, however, in any connection with the business. The entries look as though the writer had amused himself by scribbling them down, and then crossing them out again. " Miss Naples " is the only one not crossed through.

It is known that the man described as N—— by Bransby Cooper was on board the *Excellent* in the action off Cape St. Vincent. In the muster-book of the *Excellent* for 1797 Josh. Naples is down as an A.B.: he is there stated to have been born at Deptford, and to have been 21 years of age in 1795. This seems conclusively to prove that Naples was the man who wrote the Diary.

The men who composed the gang at the time the Diary was written are, in that document, nearly always spoken of by their Christian names. Their names are Ben [Crouch], Bill [Harnett], Jack [Harnett], Daniel,* Butler, Tom [Light], and Holliss. This gang, whose doings are recorded in the Diary, was the chief one in the Metropolis in the early part of the present century. The account, therefore, of the proceedings of these men gives a good idea of the work of the body-snatchers in general. Honour amongst thieves was not the motto of the resurrection-men ; they seem to have been ever ready to sell or cheat their comrades, if a favourable opportunity presented itself.

For the accompanying biographical notes of the men mentioned in the Diary the writer is indebted chiefly to the account given of them by Bransby Cooper.†

Ben Crouch, the leader of the gang, was the son of a carpenter, who worked at Guy's Hospital. He was a tall, powerful, athletic man, with coarse features, marked with the small-pox, and was well known as a prize-fighter. He used to dress in very good clothes, and wore a profusion of gold rings, and had a large bunch of seals dangling at

---

* Cannot find out his surname.   † *Loc. cit.* vol. i. *passim.*

his fob. He was tried for stealing cloth from Watling Street, but was able to successfully prove an *alibi*. Bransby Cooper states that Crouch was seldom drunk, but when he was in that state he was most abusive and domineering ; the Diary shows him in more than one of these attacks. He was sharp enough to be always sober on settling-up nights, and so had a distinct advantage over his comrades ; by this means he generally managed to get more than his proper share of the proceeds of their horrible work. About 1817 he gave up the resurrection business, and occupied himself chiefly in dealing in teeth ; in this he was joined by Jack Harnett. They obtained licences as sutlers, so that they might be allowed as camp-followers, both in France and Spain. A large supply of teeth was thus obtained by them, their plan being to draw the sound teeth of as many dead men as possible on the night after a battle. They did not limit their attention to teeth, but made large sums of money by stealing valuables from the persons of those who had fallen in battle—proceedings which were even more brutal than their former resurrectionist practices. With the money he had thus made,

Crouch built a large hotel at Margate, which at first looked like being a paying concern. The nature of his former occupation, however, leaked out, and ruined his business; he then parted with the property at a great sacrifice. Subsequently he became very poor, and, whilst Harnett was away in France, Crouch appropriated some of his property; for this he was sentenced to twelve months' imprisonment. After this he lived in London, in great poverty, and was ultimately found dead in the top room of a public-house near Tower Hill. It is very probable that at one time he made money by lending to the medical students. In his "Confessions of a Dissecting-room Porter," before alluded to, Albert Smith says, "I beg you will look at your watches, if you have not already lent them to Uncle Crouch."

Bill Harnett was a favourite with Astley Cooper and Henry Cline. With the exception of a fondness for gin, he seems to have been a more respectable man than one would have expected to find in such company. He was very obliging, and could generally be trusted

to carry out his promises. Bransby Cooper
states that Bill Harnett and " N." objected to
Crouch, and often worked against him ; in the
Diary they will be all found working together,
though there is recorded at least one " row "
with Crouch. Bill Harnett was a good boxer,
and fought Ben Crouch at Wimbledon ; he
had previously received an injury to his jaw,
and Crouch hit him a severe blow on this part,
which decided the fight in Crouch's favour.
Harnett died in St. Thomas' Hospital of
consumption. Like Southey's " Surgeon," he
had a great horror of being dissected, and on
his death-bed he obtained a promise from
Mr. Joseph Henry Green that his body should
not be opened.

Jack Harnett was a nephew of Bill ; he is
described as a stout, red-haired, ill-looking
fellow, uncouth in his address and manner of
speech. Like his partner, Crouch, he seems
to have been fond of display in the matter of
jewellery. But, unlike Crouch, he did not lose
the money he had made, and at his death left
nearly £6,000 to his family.

Butler was originally a porter in the dissecting-room at St. Thomas'. Bransby Cooper describes him as "a short, stout, good-tempered man, with a laughing eye and Sancho-Panza sort of expression." He was a clever articulator, and dealt largely in bones and teeth. Afterwards he set up as a dentist in Liverpool; but his dissolute habits were his ruin, and he was obliged to fly from his creditors. Butler was sentenced to death for robbing the Edinburgh mail, but his execution was postponed. During this delay he obtained the skeleton of a horse, and articulated it in the prison. The Austrian Archdukes John and Lewis were at that time in this kingdom, and, on visiting the prison in Edinburgh, were shown this skeleton; they were so pleased with the man's handiwork that they obtained his pardon from the Prince Regent. After his release, Butler was never heard of again by any of his old comrades or employers.

Tom Light is not mentioned by Bransby Cooper by name; he gives an account of a resurrection-man whom he calls "L———," but whether this notice refers to Light or not

cannot be definitely determined. In all probability L—— and Light are identical ; Cooper
speaks of the former as being so unreliable
that his comrades could never trust him. Tom
Light seems to have had a happy knack of
escaping justice ; on p. 92 will be found an
account of his being acquitted, even when
taken with the bodies in his possession. He
does not seem to have worked regularly with
Crouch's gang ; at Hatton Garden Police
Court he appeared as T. Light, *alias* John
Jones, *alias* Thomas Knight, in October, 1812,
and it was stated against him that he had
lately been convicted at the Middlesex Sessions
of stealing dead bodies for dissection, but he
had evaded standing his trial, in consequence
of which the Bench issued a warrant against
him. The particular charge on which he was
now brought before the magistrates was that,
with Patrick Harnell,* one of his bail, he had
been found in the act of stealing three dead
bodies from the parish burial-ground of St.

* B. Cooper gives an account of a Resurrectionist under the
name of " Patrick " ; this is probably the man referred to.
The name is Harnell in the *Sun* for October 14th, 1812 ; it
may, perhaps, be a misprint for Harnett ; two men of this
name have already been spoken of.

Pancras, or St. Giles, which were separated
only by a wall. The men were apprehended by
the horse patrol of the Hampstead and Highgate district. There was some difficulty in
carrying on the case, as, until it was determined
from which burial-ground the bodies had been
taken, it could not be said which parish was
the real prosecutor. Light attempted to
escape, but was secured. The newspaper adds,
"and, from the frequency of such offences,
strong indignation was excited in the neighbourhood, from whence a crowd attended at
the office."

Holliss was originally a sexton, and, like so
many of his class, came into the pay of the
Resurrectionists ; at last his demands became so
exorbitant that the resurrection-men refused
to pay him, and informed his employers of
what had been taking place. He was at once
dismissed, and, having no other means of
livelihood, he joined the resurrection-men.
He saved money, and afterwards purchased a
hackney coach, which he himself drove.
Like most of his companions, Holliss came to
a bad end. Harnett, the younger, had been

to France, and had brought away with him a large number of teeth, which he valued at £700; these he entrusted to his daughter, who left them in a hackney carriage. The driver found the teeth, and, not knowing how to dispose of them, consulted his friend, Holliss. Holliss offered £5 for the teeth, and promised an extra sum if they sold well. Harnett had made known his loss to Holliss, so that he knew perfectly well to whom the teeth belonged. Thinking that he could make more money by selling them privately than by trusting to a reward from Harnett, he began to dispose of the teeth to dentists. Harnett made enquiries of some of his customers as to whether they had lately been offered teeth for sale, and was shown some lately purchased from Holliss; these he was able to identify. Holliss was at once given into custody, and was tried at Croydon; he escaped transportation through a flaw in the indictment. Whilst he was in gaol awaiting his trial, Harnett seized Holliss' house and all his household furniture for a debt of £83. Holliss was afterwards mixed up in a horse-stealing case, and ultimately died in great poverty and wretchedness.

"N." or Joseph Naples, the writer of the Diary, is described by Bransby Cooper as "a civil and well conducted man, slight in person, with a pleasing expression of countenance, and of respectful manners." He was the son of a respectable stationer and bookbinder, and in early life went as a sailor into the King's service. He was for some time on board the *Excellent*, and served in that vessel in the engagement off Cape St. Vincent.* Then he returned to England, and, having spent all his prize-money, went on a vessel cruising about the Channel. From this he ran away and came back to London; here he obtained a situation as grave-digger to the Spa Fields burial-ground. A man named White enticed Naples into the resurrectionist business; this soon caused him to lose his situation. White was stopped by the patrols, and a body was found in his possession. He managed to escape, but it was proved that the body had been taken from Spa Fields, and Naples was sentenced to two years' imprisonment. He escaped, with another prisoner, from the House of Correction by making an opening through

* See also p. 126.

a skylight in the roof, and afterwards scaling the outer walls of the prison by means of a rope.

He was retaken through information given against him by Crouch, and it was only by the mediation of Sir Astley Cooper with the Secretary of State that Naples escaped additional punishment. In the list of prisoners written out by himself, and printed by Bransby Cooper (*Life*, vol. 1. p. 423), Naples thus describes himself: " Jos^h. N——-* ' Resurgam Hommo,' for trial."

The writing and spelling in the Diary show him to have been a man of superior education to most of his class. He continued in the resurrectionist business up to the time of the passing of the Anatomy Act, when he was taken on as a servant in the dissecting-room of St. Thomas' Hospital.

There is considerable difficulty in identifying many of the burial-grounds from which bodies are said in the Diary to have been stolen. Many of these were private, and the name

---

* The name is suppressed in the printed copy.

mentioned is probably either that of the proprietor or of the care-taker. No doubt, too, some of the names are slang terms which are quite forgotten at the present day.*

* Since the above was written, Mrs. Basil Holmes' interesting volume on *The Burial Grounds of London* has been published. Reference to this book confirms the statement above made. Mrs. Holmes' account is very carefully done, and the list of the old burial-grounds is probably as complete as it can be, but no light is thrown upon any of the difficult names used in the Diary.

FACSIMILE OF PAGE OF DIARY.

THE DIARY

[The spelling of the Diary has been preserved in the reprint, but as there is no attempt at punctuation in the original, stops have here been added to make some of the entries more intelligible. The writer's capital letters, too, have not been strictly followed in the reprint.]

1811 NOVEMBER.

Thursday 28th.   At night went out and got 3, Jack & me Hospital Crib,* Benjn, Danl & Bill to Harpers,† Jack & me 1 big Gates,‡ sold 1 Taunton D^o St Thomas's.

[* Slang for a burial-ground.]

[† Harper is probably the name of the keeper of a burial-ground.]

[‡ This occurs often in the Diary, and was evidently a favourite place for meeting. It was, doubtless, the entrance to some burial-ground, but there is no evidence by which the place can be definitely determined.]

Friday 29th.   At night went out and got 3, Jack, Ben & me got 2, Bethnall Green, Bill & Danl. 1 Bartholo^w. Crib opened;* whole at Barth^w.

[* *i.e.* a body which had had a *post mortem* performed on it was obtained from the burial-ground attached to St. Bartholomew's Hospital.]

Saturday 30th.   At night went and got 3 Bunhill Row, sold to Mr. Cline, S^t. Thomas's Hospital.

139

REMARKS, &c., DECEMBER, 1811.

Sunday 1st. We all look^d. out,* at Home all night.

[* Watched to see what funerals were taking place during the day.]

Monday 2nd. Met at S^t. Thomas's, Got paid for the 3 adults & settled ; met and settled with Mordecei,* made Him up £2 5s. 6d. and Receipt of all demands. At Home all night.

[* Probably Michael Mordecai, who kept an old curiosity-shop in New Alley, and was a noted receiver.]

Tuesday 3rd. Went to look out and brought the Shovils from Barthol^w., Met early in the evening at Mr. Vickers,* did not go out that night, Butler and me came home intoxsicated.

[* Probably the landlord of a public-house.]

Wednesday 4th. At night went out and got 10, whole* went to Green† and got 4, Black Crib 1, Bunner‡ fields 5.

[* *i.e.* all the gang.]

[† The "Green Churchyard" was an addition to the Churchyard of St. Giles, Cripplegate. "Green Churchyard" is a name which we find repeated in other parishes ; for instance, it was given to the higher portion of St. James', Piccadilly, and to the little piece by St. Bartholomew the Great, approached through the present south transept. HOLMES, *loc. cit.* It is impossible to say which of these is here meant.]

[‡ Bunhill.]

Thursday 5*th*. The whole at home all night.

Friday 6*th*. Removed 1 from Barthol. to Carpue.* At night went out and got 8, Dan¹. at home all night. 6 Back S¹. Lukes & 2 Big Gates : went 5 Barthol. 1 Frampton † 3 S¹. Thomas's, 3 Wilson. ‡

[* J. C. Carpue, the founder of the Dean Street Anatomical School.]
[† Dr. Frampton, of the London Hospital.]
[‡ James Wilson, of the Great Windmill Street School.]

Saturday 7*th*. At night went out & got 3 at Bunhill Row. 1 S¹. Thomas's, 2 Brookes.*

[* Joshua Brookes, founder of the Blenheim Street, or Great Marlborough Street, Anatomical School: for references to Brookes, see Index.]

Sunday 8*th*. At home all night.

Monday 9*th*. At night went out and got 4 at Bethnall Green.

Tuesday 10*th*. Intoxsicated all day : at night went out & got 5 Bunhill Row. Jack all most buried.

Wednesday 11*th*. Tom & Bill and me

L

removed * 5 from S^t. Barthol^w., 2 Wilson, 2 Brookes, 1 Bell †; in the evening got 1 Harps, ‡ went to S^t. Thomas', at home all night.

[* See page 65.]
[† Sir Charles Bell, of Great Windmill Street School.]
[‡ Abbreviation for Harpers. See p. 139.]

Thursday 12*th.* I went up to Brookes and Wilson, afterwards me Bill and Daniel went to Bethnall Green, got 2 ; Jack, Ben went got 2 large & 1 large small back S^t. Luke's,* came home, afterwards met again & went to Bunhill row got 6, 1 of them with ————†

named Mary Rolph, aged 46, Died 5th Dec. 1811.

[* Either St. Luke's Church or St. Luke's Hospital in Old Street.]
[† Words so crossed out that they cannot be deciphered ; in all probability it originally read " with their ———— throat cut."]

Friday 13*th.* At Home all day & night.

Saturday 14*th.* Went to Bartholomew took^d. two Brookes : Pack^d 4 and sent them to Edinborough, came Home to Ben^n., settled £14 6s. 2½d. each man, came home, got up at 2 me Jack & Bill went to Bunhill Row and got 3. Ben & Daniel staid at home.

Sunday 15*th.* At home all day, Got up at 3 a.m. The whole party went to Harps, got 3, Went to S^t. Thomas's.

Monday 16*th.* At home all day & night Ill.

Tuesday 17*th.* At home all day & do. night.

Wednesday 18*th.* At Home all day & do. night.
Remember me when far away

[This line is written in the same way as the names mentioned on p. 127.]

Thursday 19*th.* Went to Bartholomew, At home all night.

Friday 20*th.* Went to S^t. Thomas's, came home and went to the play, came home : at 3 A.M. got up and went to the Hospital Crib got 5 large.

Saturday 21*st.* Went to S^t. Thomas's sent 1 to Mr. Taunton,* 2 to Edinburgh, S^t. Thomas's took 6 of the above this week, came home and stopt at home all night.

[*John Taunton, founder of the City of London Truss Society, a demonstrator at Guy's Hospital under Cline, and at this time principal lecturer to the London Anatomical Society.]

Sunday 22nd. Went and look'd out, at 4 o'clock, got up, party went to Harps got 3 large and 2 small, the whole went to Barth^m.

Monday 23rd. Went for orders to Wilson and Brookes, Met Bill, Brought one to Carpue, Sent him back to bring 2 from Barthol^w. 1 for Brookes, 1 for Bell, Ben^n and Jack got 5 small at Harpers.

Tuesday 24th. At twelve at midnight a party went to Wygate got 3 small, came back and got 2 large at Newington, Came home then settled at Ben^n, Each man's share £8 16s. 8d., at home all night.

Wednesday 25th. At Home all day and night.

Thursday 26th. At Home all day and night.

Friday 27th. Went to look out, Came home met Ben and Danl at 5 o'clock, went to Harps, got 1 large and took it to Jack's house, Jack, Bill and Tom not with us, Geting drunk.

Saturday 28*th.* At 4 o'clock in the morning got up, with the whole party to Guy's and S^t. Thomas's Crib, got 6 took them to S^t. Thomas's. Came home and met at Thomas's again, pack^d. up 3 for Edinbro, took one over to Guys.

Sunday 29*th.* At home all day and night.

Monday 30*th.* Butler and Danl. took 1 large to Framton, large small to Hornig.

Tuesday 31*st.* Met at the Harty Choak,* had dispute about the horse.

[* Artichoke Public-house.]

REMARKS, &C., JANUARY, 1812.

Wednesday 1*st.* Got up at 3 in the morn-ing, the whole party went to Guys and S^t. Thomas', got 3 adults, 1 from Guy's and 2 from S^t. Thomas', took them to S^t. Thomas', came home and met again, took one of the above to Guy's, settled for the Horse £24. At home all night. Miss Naples.*

[* See page 127.]

Thursday 2*nd.* Went down to S^t. Thomas's, got paid £7 17 6 for one adult open D° not.   Came home, met by agreement at S^t. Thomas's, did not go out, Bill not there, Came home again, at home all night.

Friday 3rd.   Went to S^t. Thomas's, took the Fœtus to the London, Rec^d. 10s. 6d. Came back to S^t. Thomas's Rec^d. £4 4s. 0d., Went home, Met by agreement, Went to the Green got 5, Jack, Ben^n and me; Danl. and Bill at home, took the above 5 to Barthol^w. at home all night.

Saturday 4th.   Met at Bartholo^w., they took 4 of the above, 1 sent to Edinburgh, I went to Brookes, Carpue and Wilson for orders, Came back, at home all night.

Sunday 5th,   At home all day.   Met at 5, whole went to Newin.* got 3.   Jack and me took them to Wilson, Came home, met at 12, got 5 & 2 small at Harps, afterwards went to the Big Gates, got 3 adults, left Danl. at home, took the whole to Bartho^m.

[* Newington.]

Monday 6th Went to Barth^w., took 1 to

the London, Jack & Tom 1 to Harnige, D° 1 to S¹. Thomas's.   Came home, in all night.

Tuesday 7th.   At home all day, Tom removed 1 from the borough to Bartholom^w. fetched £2., from there took 2 to Mr. Wilson, D° to Brookes.

Wednesday 8th.   At 2 A.M. got up, the Party went to Harps, got 4 adults and 1 small, took 4 to S¹. Thomas's, Came home went to Mr. Wilson & Brookes, Dan¹. got paid £8 8 0 from Mr. Wilson I recd. 9 9 0 from Mr. Brookes, Came over to the borough, sold small for £1 10 0, Rec^d. £4 4 0 for adult, At home all night.

Thursday 9th.   Went down to S¹. Thomas's, got paid £8 8 0, 2 adults : at home all night.

Friday 10th.   Met at S¹. Thomas's, settled each man's share £12 12 0, 3 things* on hand.

[* Slang term for bodies.]

Saturday 11th, At 4 a.m. got up & went to the Hospital Crib, got 2 adults, met at

Barthol^w., packed up 2 for the Country, sold
1 at S^t. Thomas's : at home all night.

Sunday 12*th*.   At Home all day, at 11 p.m.
met & the whole went to Wygate, got 2 adults
& 2 small, afterwards went to the Green,
got 2 large & 1 large small,* Took them to
Barthol^w.

[* See page 71.]

Monday 13*th*.   Took 2 of the above to Mr.
Brookes & 1 large & 1 small to Mr. Bell,
Fœtus to Mr. Carpue, small to Mr. Framton,
Large small to Mr. Cline.   Met at 5, the party
went to Newington, 2 adults.   Took them to S^t.
Thomas's.

Tuesday 14*th*.   At 1 A.M. got up, Benⁿ.,
Bill & me went to S^t. Luke's, 2 adults ; Jack,
Danl. Big Gates, 1 large & 1 small, took them
to Barthol^w., Came home & went to S^t.
Thomas's, afterwards went to the other end of
the town for orders.   At home all night.

Wednesday 15*th*.   Went to S^t. Thomas's,
Came back, pack'd up 2 large & 1 small for
Edinburgh.   At home all night.

Thursday 16*th*. The party met at the Hartichoak. Settled the above, Each man's share £8 4s. 7½d. At home all night.

Friday 17*th*. Went & look out : came home met at 11, party except Dan¹., Went to the Hospital Crib & got 4, was stopt by the patrols, Butler, Horse & Cart were taken.

Saturday 18*th*, Went to the White horse, Butler bailed : at home all night.

Sunday 19*th*. Went & look'd out, at home all night, Could not get the horse out of the Stable.

Monday 20*th*. At home all day & night, Butler & Jack got drunk.

Tuesday 21*st*. Look'd out, Jack & Butler drunk as before, hindred us of going out. At Home.

Wednesday 22*nd*. At 4 o'clock in the morning got up, Bill & me went to the Hospital Crib and 1 for Mr. Cooper's* Lectures, had a dispute with the party, at home all night. Ben got drunk.

[* Afterwards Sir Astley Cooper.]

Thursday 23rd. Met at 10 at night, went to Wygate, got 4 large and 1 small, went to the Green got 3 large. Dan¹. not with us.

Friday 24th. Met at 11 at night. Met the patrols. Got one Hospital Crib and 6 at Bermondsey, took them to Barthol^w., sent 3 to the Country.

Saturday 25th. Met at Bartholomew. Took 1 to Mr. Carpue; S^t. Barthol^w. took 2 : at home all night.

Sunday 26th. Went to Big Gates to Look out, came home, at home all night.

Monday 27th. At 2 o'clock in the morning got up, met the party except Dan¹., Went to the Big gates, got 4 Took them to Barthol^w., Afterwards met, took 1 to Mr. Cline, 2 to Mr. Wilson, came home. Tom & Bill got drunk, did not go out.

Tuesday 28th. Went to Barthol^w., could not sell, came back to the Borough & came home, at home all night.

Wednesday 29*th.* Went to Bartho^w. brought remaining 2 to S^t. Thomas's, at home all night.

Thursday 30*th.* Went to S^t. Thomas's, at home all night.

Friday 31*st.* Went to look out, at night went out, got 2 Guys & Thomas's, same night 3 Harps 2 small : same night the Cart broke down, took 2 to Guys.

REMARKS, &C., FEB., 1812.

Saturday 1*st.* Went to Barthol^w., did not settle, at home all night.

Sunday 2*nd.* Went to look out, met at 5 in the evening, went to the Green, got 7 large & 3 small and 3 fœtus. Same night went to Wygate 4 large & 2 small. Took them to Bartholomew.

Monday 3*rd.* Went to Bartholomew, at home all night.

Tuesday 4*th.* Met at Bartholomew, settled

each man's share £21 9s. 4d., Met at night, went to Guy's got 3 adults. Took them to Bartholomew : at home all night.

Wednesday 5*th*. Went to Barthol^w. Met at night. Got 5 at Newington.

Thursday 6*th*. Went to St. Thomas's : at home all night.

Friday 7*th*. Met together me & Butler went to Newington, thing bad.* Jack & Bill went to Goswell St.† got 1. Came home met again. Danl. Bill & me went to St. Johns got 2 adults. Ben^n and Jack went to Flemish‡ got nothing, took 2 St. Thomas's.

[* Body putrid, and therefore of no use for anatomical purposes.]
[† Probably Church of St. Thomas, Charterhouse.]
[‡ The burial-ground for the parishes of St. Olave and St. John, Southwark ; it was taken by the " Greenwich Railway Company " : part of the approach to the " Flemish " now forms the approach to London Bridge Station.]

Saturday 8*th*. Went to St. Thomas's, at home all night.

Sunday 9*th*. Went to Look out, at home all night.

Monday 10*th*. Met. Went to St. James's. Got 9 large & 4 small, took them to Barthol^w.

Tuesday 11*th*. Went to Barthol^w. Moved the things. At home all night.

Wednesday 12*th*. Went to Look out, Met at night, went to St. Olives. Got 2 adults and 1 Do Bermondsey, Took them to St. Thomas's.

Thursday 13*th*. Met at St. Thomas's. At home all night.

Friday 14*th*. Met by appointment, me & Jack went to Big gate got 4, the rest went to St. Luke's got 2, took them to Barthol^w.

Saturday 15*th*. Met at Barthol^w. At home all night.

Sunday 16*th*. Went to Look out, at home all night.

Monday 17*th*. Met & went to Wiegate. Got 8 large & 1 small. Took them to St. Thomas's.

Tuesday 18*th*. Met at St. Thomas's. Took 2 over to Guy's. Came home & settled each

man's share £23 6s. 9d. On hand 2 open'd
Large, 3 small & 3 fœtus not paid, at home all
night.

Wednesday 19*th.* At Home all day sick.
John Harnet and Butler got drunk, at home
all night.

Thursday 20*th.* Met and went to Pancress*
got 15 large & 1 small took them to Bartholᵂ.

[* This is, of course, not the St. Pancras Church in the Euston
Road, but the old parish church situated on the north side of the
road leading from King's Cross to Kentish Town.]

Friday 21*st.* Met at Bartholᵂ. Sold part,
came home. Met at 2 a.m. went to St.
Thomas's Crib. Got 3 large, met the Patrols.
took 1 to St. Thomas's and 2 to Bartholᵂ.

Saturday 22*nd.* Met at Bartholᵂ. Sent 7
into the Country, distributed the rest about
town. At home all night.

Sunday 23*rd.* At Home all day and night.

Monday 24*th.* Bill Jack Tom and Benⁿ
with Nat Ure Getting drunk oblige to Come

Home in a Coach which prevented us going out to Harps.

Tuesday 25*th*. At home all day, at Night met at Jack to go to Harps. the moon at the full, could not go.*

[* See page 124.]

Wednesday 26*th*. Went to look out. Could not go out Jack and Tom got drunk. Ben[n]. taken very ill.

Thursday 27*th*. Went to St. Thomas's, sold the extremities. At night Tom & Bill got drunk at the Rockingham Arms, at Home all night.

Friday 28*th*. Met at Jacks Got 4 large 1 Small and 1 Fœtus, Harps. Took them to the London.

Saturday 29*th*. Met at St. Thomas's at home all night.

MARCH 1812.

Sunday 1*st*. Met & went to the Big gates got 3. Took them to St. Thomas's, not settled.

Monday 2nd.   Met at Mr. Vickers, Jack & Tom went to the fight, at home all night.

Tuesday 3rd.   Went to St. Thomas's, at night went to Pencress got 8 adults, 2 small and 2 fœtus.

Wednesday 4th.   Met at Jack's & settled, at home all night.

Thursday 5th.   Went to St. Thomas's ; at night early, went out & got 1 St. Thomas's Crib : at home all night.

Friday 6th.   Went to look out Big gates Green, at night got 1 Big gates.

Saturday 7th.   At Home all day and night.

Sunday 8th.   Met at Night, Jack, Tom & Dan¹. went to Harps got 5 Large, Bill and me went to the Big gates, miss^d.*

[* Failed to get a body.]

Monday 9th.   At Home all day and night.

Tuesday 10th.   Went to St. Thomas's & settled.

Wednesday 11*th*. Went to the Big Gates to Look out, at night the party went to the above Place and again miss^d., all got drunk.

Thursday 12*th*. At Home all day & night.

Friday 13*th*. Went to look out, met at night and went to Wiegate got 5 large, also went to the Green got 8 large took them Bartholomew.

Saturday 14*th*. Met at Barthol. sent 5 to Edinburgh, Mr. Wilson 3, Brookes 2, Barthol. 1. Settled each man's share £3 6 8. 2 on hand.*

[* Bodies unsold.]

Sunday 15*th*. Went to Look out, at night went to St. John's, Got 1 Large and 1 Large Small, Burnt. Took the Large to Wilson, small to St. Thomas's. Paid Hollis £11 11 0 at the order of Miss Kay.

Monday 16*th*. At Home all day went to Harps got 3 Large and 1 Large Small, 1 Small, and 1 Fœtus, took 2 Large to St.

M

Thomas's, 1 Large to Guy's, Large Small to
Mr. Frampton and 1 small to Mr. Taunton.
Mr. Frampton called at 7 in the evening.

Tuesday 17*th.*   Went to the Borough : at
Night met at 6 in the evening, went to the
Flemish, Jack, Ben[n]. & myself.   Got 2 adults,
Bill not with us, took the above 2 to St.
Thomas's.   Big gate for time is very well.

Wednesday 18*th.*   Went to the Big gates to
Look out.   came home, at home all night
which was a very bad thing for us as we
wanted some money to pay our debts to
several persons who were importunate.

Thursday 19*th.*   Met at Jack and settled
each man's share £6 18 4 : at 6 in the evening
went to the Meeting Crib*, 1 Large and 1
small.   afterwards went to the Big gates got
2 Large took them to Barth[w].

[* Probably a burial-ground attached to a meeting-house.]

Friday 20*th.*   Went to St. Thomas's, at
home all night.

Saturday 21*st.*   Jack and Tom got 2 Large

small, 1 Frampton 1 Taunton. Rec^d. for the above £3 13 6 D°. £4 4 0: at home all night.

Sunday 22*nd.* Went to the Green, at Home all night.

Monday 23*rd.* At home all night.

Tuesday 24*th.* D°.

Wednesday 25*th.* Went to Pancress got 5 adult, Took them to Bartholomew.

Thursday 26*th.* Went to Look out, Jack got 2 Large small. 1 D°. Frampton £3 13 6 1 D°. Mr. Taunton £4 4 0.

Friday 27*th.* Went to Look out, at Home all night.

Saturday 28*th.* Jack got 1 large small for Mr. Cline £4 4 0, at Home all night.

Sunday 29*th.* Went to the Green ; at home all night.

Monday 30*th.* At Home all day & night.

Tuesday 31*st.*   Went to Pancress got 5 adults Ben Bill & me.   Dan'l Jack and Tom went to Harps, missed.

Wednesday APRIL 1*st.*   Party went to the Green got 4 adults ; being the 1st of April the man left us a new Hat.*

[* The diary is torn at the margin in this place : the word "left" is probably correct, but who "the man" was cannot be determined.]

Thursday 2*nd.*   Went Bill & me to the Big Gates 2 adults and 1 small, Jack, Ben and Dan'l got 4 adults, Harps.

Friday 3*rd.*   Went to look out and distribute the above, met at Jack's at night, Ben being Drunk disappoint'd the party.

Saturday 4*th.*   Met and settled £108 13s. 7d. each man's share £18 2s. 3d., at Home all night.

Sunday 5*th.*   Went to look out met at Jack's at 10, not coming home in time did not go out.

[*No date put.*]   Tom & me went to Olives,* did not succeed.

[* St. Olave's.]

[At this point the diary leaves off abruptly : the entries from Friday 7th to Saturday 29th are in a different handwriting from the rest of the MS.]

[1812, August]

Friday 7th. Went to look out Hollis & myself could not find Bill, went to St Johns, then to White Chappell returned at night, went to White Chappel did not succeed, came back went to St. Johns, the other party had got the adult, coming back with the ladder, Bill got taken unto the Watchhouse,* with the ladder, came home.

[* Probably from information given to the police by the other party who "had got the adult."]

Saturday 8th. canines 11 shillings : went to union hall* Bill got clear the party went to Bartholomew. At Night went to W⁰. Chappel got 4 adults, one small, took 2 to Bartholᵐ. 2 & 1 Small to St. Thomas's. Horse & Cart.

[* The police court in Union Street, Southwark ; it was removed in 1845.]

Sunday 9th. at home all day & night, Wortley came concerning horse & cart.

Monday 10th. went to St. Thoˢ. got paid for 1 adult £4 4s. od. went to Bartholᵐ. got paid £4 4s. od. row'd with Ben did not settle each man had £2 2s. od, left with Hollis £2 2s. od. for Expences, at home all night.

Tuesday 11*th*. had information Crouch had cut the subjects* went to St. Thom⁵. had not cut them, Barthol^m. they had, went to differ^t. parts of the Town for orders, settled our Expence & what we had Rec^d. each man's share £3 1s. 2½d. one adult St Thom⁵. 1 London D° unpaid ½ D° Barthol^n. unpaid ; at night went to Hoxton, 1 Large Yellow Jaundice sold at Brooks.

[* *i.e.* had spoiled them for anatomical purposes ; very likely to be done out of spite, as on the previous day they had " row'd with Ben," *i.e.* Crouch ; see page 49]

Wednesday 12*th*. Went to look out, at night went to Barthol^m. Crib. cut off the extremitys took to Barthol^m.—Rec^d. £1 0s. 0d.

Thursday 13*th*. Went to St Thomas's Crib could not succeed came back to the White horse, Bill got arrested,* Millard † pd. the Debt I got drunk would not go out, Bill & Hollis went to Weigate got 1 adult male, took to the Boro, Rec^d. £2 0s. 0d.

[* Evidently for debt.]

[† Millard was superintendent of the dissecting-room at St. Thomas' ; he was an avaricious man, and lost this situation through dealing in bodies. IIis plan was to take them in at the hospital from the resurrection-men, and then to sell them at an advanced price in Edinburgh unknown to the men who supplied him, and to the teachers at the hospital. Millard was popular with the pupils, and, after his dismissal, they persuaded him to take an eating-house in the neighbourhood of St. Thomas'. As there was money to be made in the "resurrection" traffic, he did not abandon his connection with the body-snatchers. This came to be known, and created a strong prejudice against him; so much so that his legitimate business fell off to such an extent as to make it necessary for him to relinquish it altogether. Then he took entirely to the resurrection business, and was sentenced to three months' imprisonment for taking a body from the burial-ground attached to the London Hospital. He appealed against the sentence, and found bail. Then he brought an action against the magistrate at Lambeth for false imprisonment; this was set aside, and Millard was sent back to Cold Bath Fields to complete his sentence. He tried hard to get Sir Astley Cooper to solicit a pardon for him, but without avail. This so preyed on his mind that he threatened Sir Astley with bodily injury. Ultimately Millard quite lost his reason, and died in gaol. In 1825 his widow published a pamphlet entitled, "An Account of the circumstances attending the imprisonment and death of the late William Millard, formerly superintendent of the Theatre of Anatomy of St. Thomas' Hospital, Southwark." The pamphlet states that Millard had notice to leave St. Thomas' because it was found that he was supplying Mr. Grainger with bodies, and that Sir Astley Cooper was determined to put an end to the school which Grainger had established. The publication is of a very abusive character; the surgeon of the gaol, Mr. Wakley, of the *Lancet*, and the authorities at the hospital, all come in for severe censure. The whole tone of the pamphlet is so exaggerated that it is impossible to tell whether there is any truth in Mrs. Millard's grievances.]

Friday 14*th.* Went to Barthol swagg^d. the Extra^s* to St. Thos. at night went to Weigate got 1 male took them to Brooks Dan — —† Rec^d. £1 1 0

[* Extremities.]

[† These words are illegible.]

Saturday 15*th*.   Went to different places, at Night went to panc^s.* Miss^d. Rec^d. £1 o ö

[* St. Pancras.]

Sunday 16*th*.   Went to look out, at Night went to Harp's got 1 adult male took to Wilson 1 Small D^o, took to Bartho^w: a Porter carried the large.   Hollis did not go in.

Monday 17*th*.   went & got paid for the above, small fetched three guineas, at night went to Connell^s. got 1 adult M.* Danl. carried to St Thos.   Hollis did not go in, Rec^d. £1 o o.   [* Male.]

Tuesday 18*th*.   Went to different places, at night went to the play.   Rec^d. £1 10 o.

Wednesday 19*th*.   Went as above at night Flemish 1 ad^t. male, took to St Thomas's got paid; likewise head, Millard gave £1 1 o for it.

Thursday 20*th*.   As above, at night went to Flemish 1 adult male, took to St Thos. Rec^d £1 12 o.

Friday 21*st*.   Went to different places, settled our quarrell by agreeing with Mr.

Stanley* to bring in a subject for Lecture,
removed the above from St. Thom[s] at night,
went to Harp's got 1 adult M. underneath
took to St. Thos[s].

[* Mr. Edward Stanley, Surgeon to St. Bartholomew's Hospital.]

Saturday 22nd.   Went to look out me &
Hollis, Bill staid in the Boro, got paid £4 4 0
for the above a very queer one, received but
two Guineas for the one at Barth[l]. would not
do for Lecture, Sett[d]. each man's share
£1 16s. 6d., at night the party went to
Weigate, the thing bad.

Sunday 23rd.   Went to look out at different
places.   Holliss met with Ben at St. George's
agreed to meet at Lamberts with the seperate
partys : met, look[d] at each other nothing trans-
pired concerning the Business, our party went
to Harp's could not get in the private* door
being shut, came home.

[* Private door into the burial-ground, probably generally left un-
locked for them by the custodian ; for some special reason it was closed
on this particular night.]

Monday 24th.   Our party went to Barthol[m].
met with Ben and Daniel partly agreed me &

Ben went in the Cart to different places to look out coming back by Charing Cross met the Jews* Drag touted till Dark and lost scent came home did not go out that night.

[* In all probability Israel Chapman, a Jew, who was in the resurrection trade ; the object of following was to try and prevent his doing any business. (See page 49.) The next entry shows that the Jew had sold a body at St. Bartholomew's ; there was "a row" at this, and, no doubt, "the regular men" had to be pacified.]

Tuesday 25*th*.   Understood the Jew had brought a Male to Barthol^m.   Met by appointment at the above place, had a row, came home after looking out, met by agreement at 4 p.m. (Crouch having over slept himself two hours) went to St J^ns work^d. three places only got two adults M. and F. on account of not having time, took the above to Mr Frampton.

Wednesday 26*th*.   Seperated to look out. Holliss got paid in part £6 from Mr. F.   I took from the above £1. the party met at night, Ben Bill & Tom Light went to St Geo^s got 2 adults M. & F. took 1 Willson M. & F. Barthol^m- me Jack and Hollis went to Isl^n. could not succeed the dogs flew at us, afterwards went to Pancr^s found a watch* planted, came home.

[* Placed there by friends of the deceased, in all probability.]

Thursday 27*th.* Went to look out, Hollis got paid the remainder at Frampton £2 8 0. Met at night at St. Thos.—very light could not go out (came home).

Friday 28*th.* Seperated to look out, brought the F. from Bartholⁿ. to St. Thomas, having not settled took from Hollis £1 0 0, afterwards met at St. Thos. & went to St. Jns, Ben not with us work'd two holes one bad, drew the C.ⁿˢ * & took the above to St. Thos.

[* Opened two graves ; one body too decomposed to bring away, so they drew the canine teeth and sold them.]

Saturday 29*th.* Met at St Thos. could not get Pᵈ for either. Borrowed £2 of Jack, at home all night.

Sunday 30*th.* Went to look out, at night went to H. got two large Mˢ. went to St Thos. removed 1 to Wilson, 1 adult came from St Jns.

Monday 31*st.* Went to look out ——— ——— ———* old small £1 10 0 got pᵈ one do Wilson's at Night met except Bill went to Connⁿ got 2 adults M. & F. took to Bartholᵐ. one small do.

[* Words crossed out and illegible in the MS.]

1812 SEPTEMBER.

Tuesday 1st. Went to Barthol^m Got Paid for 2 adults £8 8 0, at night met, me and Light went to Connelly got 2 adults M. and one large size small F., Jack Bill & Hollis went to Weigate, got 1 large & 2 small, took 2 the above to Frampton 1 M. & D°. F. 1 large & 2 small to St Thos. 1 small to Wilson.

Wednesday 2nd. Went to the London Hollis got Canines £8 8 0, Bill got paid for 1 large M. £4 4 0. I rec^d. £4 4 0 for 1 large size small, Bill Rec^d. £1 0 0 for the F. that come from St George 1 Small came Wiegate went to Wilson. Rec^d. £2 0 0 for 1 large Small came from Wiegate, went to St Thomas' not sold being putrid : at night the party met & divided, me & Hollis went to Harp's work^d. the thing, proved to be bad, Jack Bill & Tom^s. Light went to Westminster

1812 OCTOBER.

Monday 5th. Went to look out at different places, at night party went to Lamb^h got 2 adults and 9 small took the whole to the

Borough. Mr. Cline took the about [above], 2 adults £8 8 0 from Lamb^h. & 1 small from D° £3 13 6.

Tuesday 6th. At Night did not go out.

Wednesday 7th. Went to look out Jack & Ben had a Row at the White Horse : did not go out.

Thursday 8th. Party went to see the fight did not go out.

Friday 9th. Went to look out at different places. At night went to St. Olave, got 2 adults M. & F. M. was opened took them to St. Thomas's ; again met, I got drunk, I miss^d. going with the party.

They seperated, part went to Lambeth got 1 adult F. They missed, one took that to the Boro the others (Except Ben who was getting drunk) went to Connolly got 1 adult F., took that to Bartholomew, & removed the other same place.

Saturday 10th. Met at Barthol^m. Mr. Stanley took three of the above 2 F. £8 8 0

one adult M. being opened £3 13 6d, left one
on hand, came home, in all night

[The above entry finishes a page : the back of this page is occupied
by a table for finding the moon's age on any given day : this was most
useful to the resurrection-men, as they could not work undisturbed on
moonlight nights.]

Sunday 11*th.*   Went to look out at Night
the whole party went to the Black (Jack with
us for the first time going out) got three ad^t.
M., then to Connolly two ad^t. M. & F. took
the whole to St. Thos. came home.

Monday 12*th.*   Went to St. Thos. Cline
had taken the above, went to Lamb^h. in the
evening met by appointment, at home, Drew
up an Account but did not settle Jack & Bill
not being present and others having over
drawn met again at twelve, the whole excepting
Butler went to Lambeth got 5 ad^ts. 1 Small,
Took 2 of the above & 1 small to Wilson 1 do
Carpue, 2 do Brooks, came home.

Tuesday 13*th.*   Went to Barthol^m. me Ben
Jack & Butler could not find the others,
myself came to the Boro sold 1 small that
was on hand for £1.   Came home afterwards
went to Tomlight* understood he had rec^d.

[* Tom Light.]

the money got £5 from him, came home, at home all night

Wednesday 14*th*. Went to Lamb^h. look out, at home all night on account of Ben getting out of the way, did not go out, had a dispute at Hollis's House   Child's dance.

Thursday 15*th*. Went to look out, came home went to the play, afterward met at the White horse, the party excepting Ben had a row with Hollis; seperated me, Light & Butler went Lamb^b. 2 adts, Jack, Bill & Mr. Hollis went to Connelly 5 adults, also went to St. Geo^e. 4 adts. on account of the Boy deceiving us at Lamb^h. lost the above 4 at St. Geo^e. Ben* went to France.

[* See also p. 129.]

Friday 16*th*. Met and went Wiegate got three adults 2 F. 1 M. took to Barthol^m. Jack & Tom brought the cart, came home.

Saturday 17*th*. Met at Barthol^w. rec^d. £2 0 0 went to Lamb^h. P^d. the man £1 1 0 came home, at home all night.

Sunday 18*th*. Went to look out, nothing done, at home all night.

Monday 19*th*. Went to Lamb^h. got 1 Adult M. [opened another whole but bad with the small pox]* took the above M. to Barth^m. came home, Butler not with us.

[* The words in brackets are crossed out in the MS.]

Tuesday 20*th*. Went to Barthol^w. Bill had got P^d. for the above Male I borrowed of him £1 10 0, went to Lamb^h. came home at night met at the White [Horse] Hollis myself Jack & Tom Light, Bill not with us could not find his clothes* : went to Lamb^h. two adults M. took to Barthol^w. Butler again not with us came home.

[* *i.e.* The clothes specially used for resurrection work ; they would naturally be clay-stained, and if worn during the day would betray their owner's occupation.]

Wednesday 21*st*. Went to Barthol^m got P^d. the above 2 adt. gave Light & Hollis 4s. 2d. gave Jack £2 0 0 kept £2 2 0 myself, came home, but Hollis & Light went to the Hospital Crib got 1 adult male took to St Thos. shared the money betwixt them : likewise 1 Pound for a small, at home all night.

Thursday 22*nd* went to look out, followed a black* from Tower hill, came home and met at W⁶. horse, the party except Butler went to Lamb^h. got 3 adults 2 M. 1 F. (left one behind us) 1 small & 1 Fœtus, took them to the Boro.

[* Probably slang for a funeral.]

NOVEMBER 1812.

Monday 16*th* the party went to Tottenham got 4 adults, Wilson 2. Abernethy 1. 1 on hand

Tuesday 17*th*. At home.

Wednesday 18*th*. At home.

Thursday 19*th*. Met with Hutton at Smithfield, Bill me & Ben went to St T^s got 2 ad. Jack remained with Hutton, the party went Barthol^m. C^b. got 2 ad. the whole Abernethy. Gave one to Hutton for information.

Friday 20*th*. Butler got drunk in the morning, the party except him met at Barthol^m. Me Jack & Ben went to St T^s. got 4 adt. sent Bill again after Tom to bring the Cart, took them to Barthol^m. Me Jack & Ben went to Blue Lion got 1 adt. sent Bill to bring Tom with the Cart, took that to Barthol^m.

N

came over the water got 2 adt. Guys C^b. &
1 at Thos. Crib. pack up 4 for Edinbro on the
Saturday: settled our money at home all night.

Sunday 22*nd*.   Look^d. out at St T^s. B.—
L^n*—and Tott^n at home all night.

[* Blue Lion.]

Monday 23*rd*.   Met at Barth^m. went to St
T^s. got 3 took them to Wilson, Bill took 1 ad.
to Frampton.

Tuesday 24*th*.   Went & mov^d. one of the
above to Carpue, got p^d. came home met at
Jack at 5, Bill not at home, did not go out till
morning.   Jack sold the Canines to Mr.
Thomson for 5 Guineas.

Wednesday 25*th*.   Met at Jack at 2 P.M.
Butler & myself went to the B. L^n. got 1 adt.
Jack, Ben & Bill went Panc^s. got 5 adt. & 1
small, took them to Barthol^w.   Removed 3 to
Cline, got 2 sets of can^s.*

[* Canine teeth.]

Thursday 26*th*.   Met at Barthol^m, me & Jack
went to Tottenh^m. got 1 adt.   Ben & Bill went
to St Ths. D^o 3 large came home me & Jack
got 1 Tottenh^m Bill & Ben 1 large 2 small.

Friday 27th. Met at Plough, went to St T^s. 6 adt. 1 small. Met the man with the lanthorn* : took them Barthol^m. went to Golden Lane 1 adt. 1 small gave Jack Hutton £1 as a share, took to the above place.

[* The watchman.]

Saturday 28th, Met at Barthol. disposed of the above except 1 adt. opened, 3 small, sent three to Edinboro. Drew up our Account, came home Met at Jack. did not settle, at home all night.

Sunday 29th. Went Look out at Blue L^n. &c did not go out Jack Bill & Tom Drunk the reason as Ben said for not going out.

Monday 30th. Settled our Account up to Sat^y. on hand 1 adt. Op^d. & Small three ; met at Barthol^n. me Bill & Jack Hutton went to B. Lyon got 1 adt. got up at four in the morning Tuesday, Butler Bill & me brought the above to St Thos'.

DECEMBER, 1812.

Tuesday 1st. Met at Tottenham Court Road had a dispute in St. Ts Crib. Came home did not do anything. came to the Rockingham Arms, got Drunk

Wednesday 2nd.　　Met at Vickers rectify^d. our last account, the party sent out me & Ben to St Thos. C^b. got 1 adt., Bill & Jack Guys Crib 2 adt but one of them opd. Butler look out for us, took them to St. Thos. came home Met at St Thos., me & Jack went to Tott^m. got 4 adts Ben & Bill got 6 1 1. at Pancrass took Totten^m to Wilson, Pan^s. to Barthol.*

[* *i.e.* got 6 adults, 1 small, and 1 fœtus from St. Pancras: these were taken to S. Bartholomew's: the four from Tottenham went to Mr. Wilson.]

Thursday 3rd.　　Met at Wind^ll. St. disposed of 2 of the above to Wilson, went to Barthol^w. came home for the night.

Friday 4th.　　Met at Vickers pack^d. up one for Shute, afterwards went to St Thos. got 6 adt. took them to Barth^m. left Ben & Jack Hutton to pack up for Edinbro, afterwards Jack me & Bill went to Tott^m. got 3 adt. took them to Barthol^m.

Saturday 5th.　　Remain'd at Barth^m. packing up for Edinboro, sent 12 to the wharf for the above place, at home all night.

The following list contains some of the chief sources of information on the history of the Resurrectionists.

SMITH, Southwood. "Use of the Dead to the Living." *Westminster Review*, ii., 1824, p. 59.

This was afterwards reprinted as a pamphlet. One of the editions was issued with the title of *Body-snatching*.

MACKENZIE, W. *An Appeal to the Public and to the Legislature, on the necessity of affording dead bodies to the Schools of Anatomy by legislative enactment.* 8vo. Glasgow, 1824.

GREEN, JOSEPH HENRY. *A letter to Sir Astley Cooper, on certain proceedings connected with the establishment of an Anatomical Surgical School at Guy's Hospital.* 8vo. London, 1825.

"On the Pleasures of Body-snatching." *Monthly Mag.*, iii., 1827, p. 355.

Report from the Select Committee [House of Commons] on Anatomy. Fol., London, 1828.

This is, perhaps, the best source of information respecting the Resurrectionists. Many important documents are printed in this volume, in addition to the evidence and the report.

"Importance of Dissection in Anatomy." *Westminster Review*, x., 1828, p. 116.

An Address to the members of both Houses of Parliament, on the legislative measures necessary for providing an adequate supply of human bodies for the purpose of anatomical instruction. [*By a friend of science and of man.*] 8vo. Bath. n.d.

The debates in the Houses of Parliament on the Anatomy Bills will be found in *Hansard*.

There is also much information in the pages of the *Lancet* for the period during which the Bills were before Parliament. Mr. Wakley, the editor, took a great interest in the question, and wrote many vigorous articles, pointing out defects in the Bills whilst they were under discussion.

"Supply of Subjects for Anatomy." *London Mag.*, xxiii., 1829, p. 121.

Article in *Blackwood* for March, 1829, by "Christopher North," on "Robert Knox."

The Trial of Bishop, Williams, and May, at the Old Bailey, December 2nd, 1831, for the murder of the Italian Boy. 8vo. London, 1831.

There were many Reports of this trial published, both as broadsides and as pamphlets.

"Regulation of Anatomy." *Westminster Review*, xvi., 1831, p. 482.

"Obstructions to Science of Anatomy." *Monthly Review*, cxxvii., 1831, p. 91

HANSON, N. *A Letter addressed to the Government and the Royal College of Surgeons, founded on the diabolical and horrible practice of Burking; setting forth the necessity of placing Anatomical Schools on a different footing.* 8vo. London, 1831.

GUTHRIE, G. J. *Remarks on the Anatomy Bill now before Parliament, in a letter addressed to the Right Hon. the Lord Althorp.* 8vo. London, 1832.

An Act for regulating Schools of Anatomy (2 and 3 Guil. IV. cap. 75). Fol. London, 1832.

DERMOTT, G. D. *A Lecture introductory to a course of Lectures on Anatomy, Physiology, and Surgery, delivered at the School of Medicine and Surgery, Gerrard Street, Soho.* 8vo. London, 1833.

COOPER, BRANSBY B. *The Life of Sir Astley Cooper.* 2 vols. 8vo. London, 1843.

"The Resurrectionists." *Chambers' Journal*, xxxix., 1862, p. 100.

"Body-snatching and Burking." *Once a Week*, x., 1863, p. 261.

"Burke and Hare." *All the Year Round*, xvii., 1866, p. 282.

LONSDALE, H. *A Sketch of the Life and Writings of Robert Knox, the Anatomist.* 8vo. London, 1870.

"Body-snatchers." *Every Saturday*, ix., 1870, p. 166.

FELTOE, C. L. *Memorials of John Flint South.* 12mo. London, 1884.

MACGREGOR, GEORGE. *The History of Burke and Hare, and of the Resurrectionist Times.* 8vo. Glasgow, 1884.
There is a large mass of literature relating to Burke and Hare and their trial and execution : this is well summed up in Mr. MacGregor's book.

CAMERON, SIR C. A. *History of the Royal College of Surgeons in Ireland, and of the Irish Schools of Medicine.* 8vo. Dublin, 1886.

INDEX

Plymouth: W. Brendon and Son, Printers.

www.ingramcontent.com/pod-product-compliance
Lightning Source LLC
Chambersburg PA
CBHW030538040726
47497CB00008B/2505